T0320025

Waste Worlds

Waste Worlds

INHABITING KAMPALA'S
INFRASTRUCTURES OF DISPOSABILITY

Jacob Doherty

UNIVERSITY OF CALIFORNIA PRESS

University of California Press
Oakland, California

© 2022 by Jacob Doherty

Library of Congress Cataloging-in-Publication Data

Names: Doherty, Jacob, 1984– author.
Title: Waste worlds : inhabiting Kampala's infrastructures of
    disposability / Jacob Doherty.
Other titles: Atelier (Oakland, Calif.) ; 6.
Description: Oakland, California : University of California Press,
    [2022] | Series: Atelier: ethnographic inquiry in the twenty-first
    century ; 6 | Includes bibliographical references and index.
Identifiers: LCCN 2021021910 (print) | LCCN 2021021911 (ebook) |
    ISBN 9780520380943 (cloth) | ISBN 9780520380950 (paperback) |
    ISBN 9780520380967 (epub)
Subjects: LCSH: Refuse and refuse disposal—Uganda—Kampala. |
    Urban renewal—Uganda—Kampala. | BISAC: SOCIAL SCIENCE /
    Anthropology / Cultural & Social | SOCIAL SCIENCE / Sociology /
    Urban
Classification: LCC HD4485.U332 K36 2022 (print) |
    LCC HD4485.U332 (ebook) | DDC 363.72/8096761—dc23
LC record available at https://lccn.loc.gov/2021021910
LC ebook record available at https://lccn.loc.gov/2021021911

31 30 29 28 27 26 25 24 23 22
10 9 8 7 6 5 4 3 2 1

For my parents

Contents

Illustrations

Preface

On the road that leads to Kiteezi Landfill on the outskirts of Kampala, traders have set up kiosks to deal in valuable rubbish. In the course of my research, I tried to interview as many traders as I could in the quiet periods when no one came to buy or sell the various recyclables collected at the landfill. At the end of these interviews I asked if the traders had any questions for me. One woman, a trader named Aisha originally from Eastern Uganda who specialized in cardboard, asked a question that has stuck with me: "Mwetemulina kasasiro mu nsiyamwe?" (Don't you have garbage in your country?) Indeed, we do.

I take Aisha's question as essentially ethical: Why fly so far from home to study a problem you surely have there too? What makes her available as a research subject? Will she be proud of the image of her city that will result from this work? Are Ugandans especially wasteful, or uniquely impoverished? And if not (they aren't), what makes this work meaningful? Would my questions be better posed to the world's most powerful and prolific polluters, closer to home? Doesn't there appear to be a disjuncture between the most significant causes of global pollution and climate change in the Anthropocene and ethnographic research at the peripheries of an African city?

This book is an effort to describe Kampala's waste worlds in a way that renders their inhabitants as embodying neither an exotic alterity nor a universalized abjection. Waste streams are rarely local or neatly bounded. On the contrary, they tie localities together. They are the photographic negatives of the commodity flows defining planetary urbanism. Methodological nationalism is entirely inadequate for the scales and connections of the Anthropocene. The traders, salvagers, informal waste collectors, community-based developers, volunteers, youth groups, artists, politicians, and government officials that I came to know in Kampala inhabit the same world as recyclers on the streets of Oakland, Philadelphia, and New York, the cities where much of the writing for this book has taken place. Equally, they inhabit the same world as industrial polluters from the extractive sector to the petrochemical industry to high-tech Silicon Valley chip producers. Waste here and waste there are not discrete social and environmental problems to be theorized independently. My interlocutors in Kampala are thinking and working through some of the most important problems of the environmentally devastated world we all, however unevenly, occupy. How to improve the living conditions of the planet's poorest people without relying on destructive fantasies of endless growth? How to build inclusive cities? How to clean these cities without perpetuating the violence of displacement? They are working through these problems, but they are not the problems. I hope *Waste Worlds* conveys something of what I have learned with and from (and sometimes against) their thinking.

I first arrived in Uganda in 2010 interested in the oil economy, urbanization, and the politics of work in a transnational industry that has little space for local labor despite the amount of capital involved. I moved away from this project because the pace of oil development appeared slower than even the pace of a PhD project, and because of the ethical risks that would be involved in talking to displaced people in the context of speculative and militarized land-grabbing in Uganda's oil producing regions. But while I was in Kampala conducting preliminary research, I found that environmentalists and others I was meeting were always returning to the theme of the city's garbage. Pointing to rubbish at the side of the road as we walked from his office, one environmental activist told me that these heaps of garbage are evidence of the political failures of the state and a dangerous harbinger of how badly oil extraction could go. But these heaps' meanings were not stable. Another man on another walk together told me that the trash littering the roadside was evidence of the moral failure of the population to live properly and take care of their city and their environment. The recurrence of the theme, and the contradictory ways in which Kampalans treated it, sparked my interest, especially as it articulated with my own past dumpster diving and politicizing food waste as part of Food Not Bombs activism in Virginia. In the coming year, Kampala's municipal government itself brought waste to the forefront of the city's politics in a campaign to remake urban politics and the visual environment. So I started to track the ways that ideas about waste emerged in the city, how global environmental discourses were taking shape there, and how Kampala's residents differentially engaged with and crafted lives from the city's discards.

Like many ethnographers seeking to study global processes, my research had to be multi-sited. In my case, however, this did not mean that I traveled far beyond the boundaries of Kampala. Rather, I was interested in how the city itself becomes multiple and fragmented and in the various seams and streams that unite and cut across these fragments. One of the first, and methodologically foundational, things I learned about Kampala's waste infrastructure was its ephemeral nature. Waste management organizations existed transiently and struggled to sustain themselves. By the time I arrived to do my research proper, two of the groups I had intended to study after my preliminary research visits were no longer present. One had relocated its projects to a smaller town in western Uganda, and the other collapsed when they lost access to the land they had been using. Other groups I came to know had a fleeting existence, coming together for intense afternoons or weekends of work, before members returned to the other activities that occupied their lives and earned their livelihoods. The municipal government had a more established and continuous presence (the everyday construction of which is described in part 1), but although municipal officials were welcoming of my requests for interviews and availed themselves generously when I asked to shadow them on their day-to-day routines, they were reluctant to permit me a more embedded role. Kampala's waste infrastructure thus offered no simple site in which I could achieve classic ethnographic immersion. Instead, it forced me to track difference and connection across diverse worlds of waste and projects of urban cleaning.

Ultimately, the sites I studied included elite and transnational nongovernmental organizations (NGOs) engaged in community cleaning exercises and awareness-raising programs, municipal government offices dedicated to planning and carrying out routine waste management, private garbage collection companies trying to convince the emergent middle class to pay for waste disposal, youth groups launching similar private waste collection programs as social enterprises, informal recycling economies salvaging value from plastic bottles and other resalable materials with markets as far flung as China, local politicians attempting to garner votes by delivering waste collection to their constituents, and community-based development organizations struggling to attract support from ever-elusive international donors.

My research took me to all five of Kampala's administrative divisions, as well as to the surrounding suburbs and peri-urban areas, like the municipal landfill, that are gradually being encompassed into a metropolitan region. This heterogeneity of sites accrued through snowballing. Attending one-off cleaning exercises in new neighborhoods, I often met local organizers and entrepreneurs who invited me to return to visit their projects and observe quotidian routines of waste management. Shadowing municipal workers, I encountered new dump sites and clusters of recycling businesses and made connections with workers at different levels of the municipal hierarchy. This meant that although my fieldwork did not have a predictable weekly rhythm, I was able to expand the breadth of research by following the waste stream where it led me. Some groups and individuals were especially generous with their time and patient with my questions, and they became key interlocutors with whom I sustained ongoing relationships for the duration of my fieldwork (and in several cases beyond, through social media). These relationships allowed me to observe the struggles and achievements of community organizations as they developed over nearly two years and to track small changes in the economies, legalities, and spatialities of various modes of waste work. Other groups, locations, and individuals—especially private waste companies and informal recycling entrepreneurs—welcomed me for single visits to conduct focused interviews. These allowed me to pose more standardized sets of questions and to observe commonalities and diversity across the city. I thus matched different interview techniques to different sites (e.g., daylong unstructured conversations as I shadowed a municipal waste supervisor, highly targeted interview questionnaires for busy waste company operations managers, and group interviews with plastic salvagers at trading kiosks) and paired these with participant observation, documentary photography, and media analysis to gather and analyze ethnographic data. Kampala's lively press also emerged as a critical source not only of information about urban policy and politics but also as a site in and of itself where waste and waste infrastructure were represented, discussed, mediated, and thus made public in important ways. I draw on planning archives, policy documents, and environmental gray literature (focusing primarily on the period from 1986 to the present, although they also included materials from the 1960s and 1970s) available at the library of the National Environmental Management Authority and at various NGO and municipal offices to situate my ethnographic research within the multiple histories of development intervention in President Yoweri Museveni's Uganda.

The tense of my ethnographic writing moves between past and present. I use the past tense to tell specific stories and recount particular historical events and the present tense to describe infrastructures and practices that are, at the time of writing, ongoing. This research was conducted in Kampala from September 2012 to July 2014, with preliminary research visits in the summers of 2010 and 2011 and a return visit in the summer of 2018. Although the use of the ethnographic present can be a technique of objectification, distancing, and othering, my intention in selectively using the present tense here is not to offer a description of timeless practices but to acknowledge the often precarious and hard-fought continuity of many of the practices, economies, and institutions that I describe. In the context of dramatic urban change, the worlds I describe are precarious and under pressure such that to shift into a past-tense description would be to further marginalize them by completing their disposal from the present. The ethnographic present is thus intended to signal this ongoing historical presence.

ACKNOWLEDGMENTS

This book has been a long time in the making. It owes its existence to the innumerable conversations that have taught me about waste worlds and helped me think about them in new ways. First and foremost, I'm forever indebted to the Kampalans who, though they largely remain unnamed or pseudonymous in this text, made the time to talk to me during my research, shared their insights into and experiences of waste worlds, and helped me learn the right questions to ask. While I can never fully reciprocate it, I hope I can emulate the hospitality, patience, curiosity, and eagerness that was shared with me. I'm grateful to the faculty, students, and staff at the Makerere Institute for Social Research for being such intellectually stimulating hosts in Kampala, and whose library was the perfect space to read, write, and reflect in the midst of research. I also thank the Uganda National Council for Science and Technology for granting me permission to conduct fieldwork in Kampala (UNCST SS 2606). Two research assistants, George Mpanga and Oliviera Nalwanga, contributed immensely to the completion of this project. I am indebted to their skills as interviewers, their fluency in multiple languages, their willingness to approach some truly smelly places, their curiosity, and their understandings of Kampala life, urban geography, and the basics of sociality. My time in Kampala was enlivened by the friendship and camaraderie of Osman Salad America, Miriam Bina, Charles Golooba, Reagan Kandole, Philip Kalinda, and Denis Luteeka, whose collective wisdom and humor taught me so much about the city.

This book grows out of the dissertation I completed in the Department of Anthropology at Stanford University, where I was fortunate to learn from a brilliant community of teachers and students. I was privileged to work with an amazing committee whose research and writing have inspired me, whose questions have pushed me, and whose support has been immense through the life of this project. Jim Ferguson's work has been a model and a constant point of reference. I am indebted to his keen readings from the earliest days of this project that always get directly to the heart of matters to pull out nascent ideas, suggest new conceptual frameworks, identify latent debates, and situate the stakes of arguments. I am grateful for conversations with Paulla Ebron about postcolonial cities, ethnographic theory, and research methods, for her always-on-the-mark suggestions of what to read next, and for her writing that, since I first encountered it as an undergrad, has shown me how to think about global connections through the details of gesture and performance. Lochlann Jain's ethnographic creativity and empathetic attention to the material politics of injury and difference has been a genuine inspiration. I thank her too for some of the hardest questions I have been asked and for incredibly attentive multi-scalar readings that have improved my chapters and my sentences. I am grateful to have had the opportunity to learn from Jake Kosek's brilliant scholarly combination of conceptual breadth, rigor, and originality. At Stanford and beyond, this project benefited from the guidance of Thomas Blom-Hansen, Miyako Inoue, Matthew Kohrman, Helga Leitner, Liisa Malkki, Ananya Roy, Eric Sheppard, and Sylvia Yanagisako. I couldn't have asked for more dynamic and engaging cohorts to learn, think, argue, write, and laugh with than Jess Auerbach, Maria Balfer, Stef Bautista, Firat Bozcali, Samil Can, Hilary Chart, Damien Droney, Patrick Gallagher, Mark Gardiner, Maron Greenleaf, Yasemin Ipek, Eda Pepi, Elly Power, Jenna Rice, Johanna Richlin, Karem Said, and Anna West.

Since leaving California, work on the book has taken place in dozens of offices, libraries, desks, coffee shops, and couches over a peripatetic five years. I'm grateful to J. Kehaulani Kauanui, Jennifer Tucker, Gina Athena Ulysse, and Margot Weiss for the warm welcome to Wesleyan University. I also thank the students in my courses Pure Filth, Anthropology of Infrastructure, and Global Africa, who helped me think in new ways about what anthropological writing can be and do. Much of my writing was completed while I was a Mellon postdoctoral fellow at the Wolf Humanities Center at the University of Pennsylvania. I'm grateful to the center's director, James English, and the convener of the Afterlives forum, Emily Wilson, for their support, as well as to the other 2017–18 fellows for their insight and collegiality. In my two years at the University of Oxford I was challenged to rethink urban life in new ways big and small by my colleagues on the PEAK Urban project. I'm grateful to Tim Schwanen and Michael Keith for all I've learned and for fostering these lively ongoing conversations on urban futures and to everyone in Oxford's Transport Studies Unit for their warmth and openness through profoundly strange times. Final work has been completed at the University of Edinburgh, where I'm deeply grateful to Lotte Hoek, Sarah Parry, and all my new colleagues in Social Anthropology and in Sustainable Development for their guidance and unflappability, which have made this a real home throughout the most disorienting couple of years.

Intellectual support for this project from teachers, mentors, colleagues, and friends has come in myriad forms from conversations spanning years to fleeting discussions. I'm grateful to Amiel Bize, Brooke Bocast, Kate Brown, Emily Brownell, Waqas Butt, Sanda Calkins, Serena Cruz, Kevin Donovan, Sam Dubal, Henrick Ernstson, Rosalind Fredericks, Eric Gable, Claudia Gastrow, David Giles, Zack Haber, Basil Ibrahim, Jason James, Maria John, John Kuhn, Shuaib Lwasa, Rhea Rahman, Josh Reno, Farhang Rouhani, Melanie Samson, Jonathan Silver, Marisa Solomon, Joanna Steinhardt, Stuart Strange, Miriam Ticktin, Orchid Tierney, António Tomás, Vasiliki Touhouliotis, Antina von Schnitzler, Delia Wendel, Hylton White, Ara Wilson, and Tyler Zoanni, who have all contributed in ways large and small to this book. I'm especially thankful for the friendship of Marissa Mika and Erin Moore, whose ideas, projects, and analyses have deeply enriched my own over many hours in conversation in Kampala, California, and elsewhere.

I thank Anna Tsing, Nils Bubandt, Andrew Mathews, and Danilyn Rutherford for the invitation to participate in the Wenner-Gren Symposium "Patchy Anthropocene" and to the other participants for their engagement with my work. I likewise thank Ramah McKay and the audience at the University of Pennsylvania's History and Sociology of Science Workshop for a stimulating discussion of animal infrastructures. Portions of this book have appeared previously in "Capitalizing Community: Waste, Wealth, and (Im)material Labor in a Kampala Slum," *International Labor and Working-Class History*, 2019; "Labor Laid Waste: An Introduction to the Special Issue," *International Labor and Working-Class History*, 2019; "Maintenance Space: The Political Authority of Garbage in Kampala, Uganda," *Current Anthropology*, 2019; and "Filthy Flourishing: Para-Sites, Animal Infrastructure, and the Waste Frontier in Kampala," *Current Anthropology*, 2019. I would like to thank the editors and reviewers of these articles for their feedback, which helped clarify my arguments, and the publishers for the permissions to include this material here.

At the University of California Press, I'm extremely grateful to Kate Marshall and Enrique Ochoa-Kaup for the attention, care, and patience they've given this project over its long gestation. I'm also indebted to Kevin O'Neill for the faith he's shown, the encouragement he's given, and the support he's offered me over the years and for including this book in the Atelier series. Many thanks are due to Danny Hoffman and Sophia Stamatopoulou-Robbins for their thoughtful and generous reviews of the manuscript, which have improved the book immeasurably. I'd also like to thank Reed Malcolm, Jatin Dua, Antony Fontes, and Kathryn Mariner for their comments on early drafts of several chapters at the inaugural Atelier workshop in Denver and Athena Lakri for her careful copyediting of the manuscript. Thanks also to Duncan Senkumba, whose work conveys the creativity of waste worlds and brings the book's cover to life.

I wish to thank the institutions that have generously provided financial support for this project: the Social Science Research Council (Dissertation Proposal Development Fellowship), National Science Foundation (Doctoral Dissertation Improvement Grant), Wenner-Gren Foundation (Dissertation Fieldwork Grant and Engaged Anthropology Grant), Stanford University (Jeanne and Bud Milligan Fellowship), Stanford's Department of Anthropology (Summer Language Training Grant), the Center for African Studies at Stanford (Summer Language Fellowship), the Stanford Humanities Center (Mellon Foundation Dissertation Fellowship), and the Wolf Humanities Center at the University of Pennsylvania (Mellon Postdoctoral Fellowship in the Humanities).

Finally, I thank my family. I'm grateful to parents, Steve and Moira Doherty, for so much but especially for teaching me to read, write, and approach the world with curiosity. I'm thankful for my brother Josh whose thoughtfulness and commitment to justice are an inspiration. I can't imagine having completed this project without Julienne Obadia, who has been a true partner, no matter the miles between us, at every step. I learn so much from you; thank you for your love, laughter, intelligence, patience, and indignation, basically, for being the best.

Introduction

DISPOSABILITY'S INFRASTRUCTURE

I. THE STENCH OF POLITICS

"Can we retain the glory of our market?" Fred Kidamba was agitated. We were sitting in his office at Nakasero Market in downtown Kampala, which, Fred was explaining, is Uganda's oldest, proudest, and most glorious market. "We've grown up here. We love our market. It is known all over the world as the oldest in Uganda," he boasted. "Nakasero was put up by colonialists to buy food products and other goods easily. The prices were too high for many locals." The Traders' Association has tried to keep up high standards ever since. "We are the heart of the Kampala food trade. The market runs twenty-four hours: once these people leave, others come for the night." Fred was the chairman of the market's Traders' Association, charged with managing disputes between traders, maintaining the space of the market, and liaising between the market and the municipal government. I had come to ask what I had imagined were some fairly straightforward questions about waste management at the market: How much waste is generated daily? What composes it? Who collects it, and how often? Municipal policy documents I had consulted identified market waste as one of the biggest challenges to garbage collection in the city, so

I wanted to learn more. Answering these questions, it turned out, was anything but straightforward.

Fred's office was on the second floor of the main market building. Dark and stuffy, it was crowded with furniture—desks and armchairs too big for the room. The walls were plastered with bright yellow posters, left over from the 2011 elections two years previous, advertising the Traders' Association's support for President Museveni and the ruling National Resistance Movement. A five-foot-tall cardboard cutout of Museveni occupied pride of place next to a bookshelf laden with binders and newspapers. Outside the office, a balcony overlooked the market proper, with its bustling trade in fruits, vegetables, meat, and spices spilling out from the inside of the bright-red, one-story, colonial-era covered market onto a square covered with plastic tarps and shaded by colorful umbrellas. Traders jostled for space in the crowded downtown with small retail outfits touting hardware, plumbing supplies, and electrical equipment. Casting a shadow that sheltered the market from the afternoon sun, newly built arcades—shopping centers five to six stories tall—towered over the scene. These arcades, Fred feared, were the future of Nakasero. "These investors are given tenders to manage the market, but really they just want to remove us and build some commercial complex like those ones," Fred explained, gesturing to the high-rises around us. He pointed to another building across the market. "You must talk to those ones in there. They are handling our rubbish now." The office he meant belonged to the Kampala Capital City Authority (KCCA)—the newly, and controversially, created municipal authority mandated to transform Kampala's government, infrastructure, and economy.

"But you know, we have our own plan for development." Opening a binder from the shelf, he presented me with a stack of documents and unfolded a large printout of an architectural rendering of the future of Nakasero Market. The Traders' Association had commissioned this work, a proposal for a new high-rise that would house the food market, with all the existing traders in place, on the ground floor, and also included a shopping mall, hotel, and a parking garage. "It will be like a supermarket that is owned by many people. It doesn't have to be like your [American] supermarkets, it can be an African idea! We can provide fresh and organic food on a daily basis. And we want to include a small museum to remember how it has been. It is only that the KCCA will not allow us to manage our own development!"

"Now those ones are managing us, but they want us gone as well! Our traders are not happy." Fred told me that since taking over the market, the KCCA cut garbage collection from four to two times daily, leaving an unwieldy heap of rubbish to accumulate and overflow its designated place in a skip (the preferred term for a dumpster in Uganda) at the corner of the market. "Now the market is stinking so much people refuse to come," Fred complained. He continued, "People passing by see Nakasero as a stinking place, but they don't know the real reason." Fred saw the heap of trash in the market as part of an ongoing struggle to keep the market in place. Stench was a weapon produced by the municipal government to turn the public against Nakasero, he argued, so that there would be no opposition when they decided to redevelop the space and evict the traders. A savvy member of political society, Fred had reached out to President Museveni directly to protect the Traders' Association. "He is our supporter, and we are his," Fred told me. "He is very observant of issues where many people are affected; you can't remove people from the president." "But," I interjected, "wasn't it Museveni who brought the KCCA in the first place?" Smiling wryly, Fred cautioned me, "Now you are getting into politics when you said you just wanted to know about our rubbish."

This book is a study of the dynamics of development and disposability in contemporary Kampala, Uganda. It asks how people, places, and things become disposable and how conditions of disposability are challenged and undone. I explore these questions through an ethnography of Kampala's waste worlds: the official and unofficial infrastructures and economies that constitute the city's waste stream and develop around it. My conversation with Fred tracks the theoretical contours of this project: capital-led urban transformations and the displacements they engender, the materiality and affective power of garbage, the forms of labor that go into creating a clean city, the pride and vision that inform and sustain the creation of popular infrastructures, as well as the contradictions and heterogeneity of the state as it governs waste worlds.

Situated on the shores of Lake Victoria, Kampala has grown from a leafy city of seven hills to a sprawling urban agglomeration stretching well beyond its official boundaries. Depending on where and when it's measured, the city's population is between one million and four million people and is said to double and halve everyday as commuters come into Uganda's capital and economic hub to work, trade, study, pray, seek medical care, shop, and visit. The Central Business District is the heart of Uganda's economy and government, where banks, bureaucracies, courts, hotels, and transport hubs channel flows of people, power, money, and commodities through the country. New suburbs, both wealthy and impoverished, are continually under construction as glamorous shopping malls, hotels, and gated communities mushroom within the commuter belt in parallel to the more modest, low-density, auto-construction projects undertaken by the urban majority that extend the urban fabric well beyond the paved road network, sewerage system, and electrical grid. Hugely cosmopolitan and proudly traditional, Kampala hosts a vast network of international NGOs and embassies as well as multiple communities of refugees and migrants from around the region. The latter have long made homes as guests of the Buganda Kingdom, which made this area its capital as successive kings built their palaces on the hilltops that now make up the city. Between the hills are rivers and wetlands that flow into Lake Victoria and are increasingly under pressure from industrial and residential construction.

All of this booming creates waste by the ton. Industrial, construction, medical, and residential waste pose continual challenges to Kampala's ambitious and overextended municipal government. Questions about waste infrastructure in Kampala are highly charged. They cut to the core of debates over the future of the city and struggles over urban belonging: Who will and who will not have a place as the city undergoes an ambitious program of infrastructural and political transformation? Who has the moral authority to clean Kampala, and on what basis? Who will bear the brunt of pollution, and who will be cleansed from the city? These politics of cleanliness are both moral and material. Moral values are embodied in, reproduced through, and transformed by surprisingly mundane material infrastructures like dump sites, skips, trash fires, drainage channels, and garbage trucks. As Fred Kidamba feared, garbage has the ability to pollute perceptions of the spaces where it accrues and of the people who inhabit them. Garbage can produce a stench that attracts the moral condemnation of neglect and irresponsibility and, in turn, paves the way for displacement. In this, and in many other ways, waste *worlds*: it participates in the creation and transformation of urban life.

Kampala is hardly the only city grappling with waste, infrastructure, and social inequality. These are pressing questions as urbanization intensifies around the world in what many observers have called the urban century: an epoch in which the majority of the world's population live in cities and the issues facing urban populations are critical challenges for the sustainable development agenda.[1] Cities of the global south experience critical disjunctures between growing populations and limited opportunities for formal work, between increasingly visible and mass-mediated forms of extreme wealth and extreme poverty, between consumer cultures dependent on disposability and moral projects to clean and green cities, and between discourses of sustainability and extractive political economies predicated on growth. These disjunctures have provoked crises for both urban government and urban theory because the form that the cities of the global south are taking, the economies that shape them, and the political frameworks governing them little resemble the normative developmental models and canonical case studies upon which the urban theory of the nineteenth and twentieth centuries was built.[2]

Garbage and garbage infrastructure provide a unique vantage point on the urban crises of inequality, governance, and ecology because of the ways they link the most intimate spheres of social reproduction to transnational economies and to large-scale projects of urban development. Most of Kampala's municipal waste, for example, is organic matter, the byproduct of food preparation. It is generated in the city's kitchens, where the highly gendered and morally charged work of feeding families is performed. Waste infrastructure attempts to reach into and act upon these domestic spaces, linking them to new planning initiatives, drainage projects funded by the World Bank, international NGO campaigns, and more. Following the city's diverse waste streams shows how these multi-scalar connections are made in practice, how multiple moral logics, ideologies, and structures of feeling emerge and interact, and how actors in different locations imagine each other and imagine the future of the city.

Waste worlds take shape in the context of complicated socio-technical waste infrastructures whose materiality has emerged over time, sedimenting colonial racial hierarchies, morally charged ideals of respectability, nationalist developmental aspirations, and globally circulating visions of urban futures. Waste worlds are densely inhabited. They emerge through the process of inhabitation—the everyday world-making practices of cleaning, sorting, discarding, and salvaging. Kampalans' work with waste does not simply result in disposability—although it often has; it also affords people across the city's class structure an opportunity to define what a clean Kampala would mean, to assert their sense of belonging in the city, and to engage in novel practices of urban citizenship. Far from being peripheral to the real locations of politics or unproductive sites of bare suffering, waste worlds are creative and generative. Waste economies and the moral project of cleaning, in other words, have both produced disposability and offered a way to challenge and undo it.

II. WASTE IS A VERB

This book is not about waste in itself precisely, because there can be no such thing. Waste is not *an* object. Waste does not a have a singular materiality; anything can become waste. Waste is something we *do*. We waste things, money, time, and effort. Through the arts of care and repair, waste is also something we undo. What counts as waste is a moral question. Waste is a verb. It is a process, a practice, and a social category. Waste is value's co-constitutive other. Waste materializes: it takes particular material forms as a result of diverse processes of wasting. Waste poses moral questions of responsibility across class, racial, gender, generational, national, and species borders. As matter, waste can also permeate and muddy these borders, requiring cleanups to reproduce and shore up social differences and hierarchies. Because waste cannot be assumed in advance, *Waste Worlds* asks how people and places, forms of life and life forms, things and ideas *become waste*, and what worlds this *becoming waste*, in turn, constitutes.

This understanding of waste builds on Mary Douglas's foundational insights that dirt is matter out of place, the outcome of ordering, and as such a fundamentally relational category.[3] This means continued attention to the processes of boundary making, by which places are made and things become dirt, as well as to the consequences of this coding. But dirt is not the same thing—materially or conceptually—as waste. Rather, waste management infrastructures are precisely about constructing a network of places and flows that ensures that discards remain manageable waste and don't become dangerous and polluting dirt.[4] Nonetheless, as *Waste Worlds* illustrates, this process of taming and controlling the material presence and symbolic charge of garbage through waste management is always ongoing and never fully complete. Rather than reducing dirt to the ambiguities of social structure and symbolism, I focus on the infrastructures (themselves material-discursive phenomena) through which entities are contingently and temporarily stabilized as waste and the categories of cleanliness are maintained. This approach means going well beyond Douglas's framework.

In contrast to Douglas's arch-constructivist symbolic approach to waste, new materialist approaches have argued that treating pollution merely as a human cultural construct is both analytically and ethically inadequate. This critique has become increasingly urgent in the face of industrial pollution, environmental racism, soil degradation, and nuclear fallout.[5] The social and ecological crises of the Anthropocene have given new stakes to long-standing philosophical questions about the place of nonhumans, from living animals to inanimate matter, in Western political ontologies that have tended to see materiality simply as a limit to human agency. Indeed, waste figures prominently in efforts to bring matter into political theory. In the opening pages of her influential book *Vibrant Matter*, for instance, philosopher Jane Bennett encounters some litter in a drain in Baltimore. Seeing it almost as a sculptural installation, Bennett is repelled and attracted, entering into a discussion of singularity and semiotics.[6] But she does not ask who threw the litter away or, before them, who made it; nor does she wonder why the streets of Baltimore are littered. Instead, she attributes these projections to the force of things, their autonomous power to act in the world, and uses this as evidence to argue for an anti-anthropological approach to matter.[7]

Understanding waste as a verb—a historicized process of relational becoming—offers an alternative both to anthropocentric liberal-humanist views of inert matter and to anthropomorphic vitalist views of agential matter. Matter exceeds discursive corralling, practical enactment, and the social relations that discipline it, but any examination of the materiality of waste must also include an anthropology.[8] Understanding wastes' materialities requires accounting for the ways in which race, class, gender, nation, and citizenship—as well as the histories of normative regimes like biomedicine, public health, and environmentalism—simultaneously participate in the materialization of waste and are themselves materialized and sedimented through continued encounters with that matter that is cast off in the construction of social worlds. In this light, it is necessary to account for the materiality of wastes, their particular effects, the ways in which they are regulated and distributed, the places they concentrate, their movement through ecosystems and bodies, and their role in the materialization of systematic forms of structural violence and social abandonment. The material properties and relational liveliness of waste streams matter, but their materiality is neither ahistorical nor independent of the moral worlds in which they emerge.

III. DISPOSABILITY: BECOMING WASTE

The idea of disposability comes from the world of consumer goods, where it refers to one-use products, designed to be discarded. Cheap enough to throw away, these goods have been integrally linked to the remaking of production and consumption, gender and subjectivity, domesticity and urbanization, infrastructure and environmentalism.[9] Disposable goods proliferated in the United States alongside suburbs and supermarkets in the decade of post–World War II economic growth that saw the baby boom and the rise of modern American consumer culture. Made possible by new plastic technologies and cheap sources of oil, disposables promised cleanliness, freedom, and convenience.[10] This kind of disposability is an effect of material abundance, the idea that it is easier, cheaper, and more desirable to make new things rather than repair and maintain existing ones. But the public had to be taught to embrace disposability. To overcome deeply held cultural and moral objections to wastefulness rooted in the Protestant ethic, American manufacturers and advertisers framed disposability as a means to hygiene and efficiency: waste in the service of cleanliness and deliverance from domestic drudgery.[11] Marketing for disposables promised a whole new way of life, easy living through disposability—a new habitus for a new mode of mass consumption. One of the earliest disposable goods, sanitary pads, for instance, promised women newfound freedoms and mobility, interpellating new gendered subjects such as "the modern college and business woman," able to overcome bodily difference to enter the workplace.[12] Disposability has engendered new modes of subjectivity, transforming not only how we relate to the material world but who and how we are.

Disposables have become indispensable. In a modern hospital setting, for example, single-use gloves, syringes, gowns, masks, and surgical drills are vital disposable technologies intended to guarantee a sterile environment. The COVID-19 pandemic has made clear the extent to which hygiene is predicated on disposability. Shortages of disposable personal protective equipment has also illustrated the ways that shortages of disposable goods unevenly transfer the risks of disease onto vulnerable populations and essential workers, who themselves become disposable.[13] The scale of medical discards generated by the pandemic also highlights the relational qualities of waste, as cleanliness in one domain generates vast quantities of garbage elsewhere. This fact has been rendered invisible—at least to wealthier and whiter communities—through the work of routine municipal infrastructure and waste management. As Josh Reno argues, this ferrying away of waste underpins and sustains ordinary domestic life.[14] At the same time, these decades-long shifts in material culture have encountered resistance and critique from moralists and environmentalists denouncing the new regimes of wastefulness.[15]

Disposable commodities are highly visible everyday artifacts that implicate consumers directly in a mounting garbage crisis and large-scale environmental matters of concern, such as global warming and ocean pollution. As such, they have been recurrent targets for environmental action and the cultivation of novel environmentalist subjectivities, such as the emergence of ultra-austere zero-waste lifestyles.[16] Objects like the polystyrene foam cup and the disposable diaper have become touchstones for environmentalism, although this has more to do with their visibility as litter than their actual status in the waste stream.[17] Of course, disposable goods are not unique to consumer cultures of the global north; they have proliferated globally.[18] Disposables have merged with diverse vernacular practices of reuse and repair, and in so doing, have generated massive economies of salvaging and recycling that are often both vital livelihood strategies as well as massively injurious to bodies and environments.[19] New politics of repair contest disposability by demanding consumers' right to open, know, and fix consumer goods to break the cycle of planned obsolescence.[20] Similarly, by salvaging and serving discarded food, movements like Food Not Bombs critique the wastefulness of contemporary commodified food systems that render edible food disposable.[21] Everyday practices of discarding that place usable goods alongside, rather than in, the waste stream to allow for the possibility for salvage and reuse contest disposability by enacting a parallel infrastructure of reciprocity.[22]

Research in discard studies teaches us how to see individual disposable products as artifacts of regimes of disposability.[23] Throw-away culture is not a moral failure of individual participants in contemporary consumerism but a material fact about the world, something that is engineered into objects and environments. Not just located in commodities, however, disposability has an extensive spatial form, relying on specific infrastructures that manage visibility and distribute harm.[24] As environmental justice researchers in the United States have repeatedly illustrated, these infrastructures of disposability reproduce geographies of economic and racialized inequality, as injurious installations like landfills, incinerators, and chemical plants are sited in and around poor, Black, and immigrant communities. Disposability is embodied not just in the new kinds of everyday consumer habitus that disposable commodities enable but also in the ways disposability's infrastructures channel the toxic externalities of disposability into bodies and environments.[25] These material flows are also discursive phenomena guided by colonial environmental imaginaries that construct particular places and people as always-already wasted and wasteful and thus available as repositories for new waste streams.[26]

Disposability is an infrastructurally produced material condition with a distinct temporality. Made from carbons produced over thousands of years ago from ancient life forms, used in a fleeting moment of disposability, and then remaining in the environment without decomposing for thousands of years, disposable commodities are artifacts of fossil-fueled capitalism that will endure long beyond their utility.[27] Disposability thus refers to the future state of objects as waste. Disposables are here today and gone tomorrow. Disposability is thus a potentiality, combining immediate utility with the ability to be thrown away. Socially, this potentiality means disposability becomes an omnipresent fact of life, something fleeting, often felt looming, perceptible as the threat of pending displacement. Disposability can also be made present and made material in the slow unfolding of environmental violence, evidenced in uncollected garbage, in floods, in outbreaks of water-borne disease, or in toxic body burdens. Moreover, as the afterlives of waste reveal, disposable goods, people, and places do not simply disappear to "away." The time after disposal is not empty. It is a time not only of ecological injury but also economic extraction and social life.[28] As this book illustrates, waste worlds are densely inhabited, despite their disposability.

Many social theorists have picked up on the idea of disposability to critique contemporary global inequalities ranging from austerity to drone warfare, structural unemployment to refugee crises, and slum clearances to ecocide.[29] This analysis often proceeds metaphorically, arguing that people are treated *like* garbage. For example, in their book *Disposable Futures*, Brad Evans and Henry Giroux write that neoliberalism has produced "a strong tendency to view the vast majority of society as dead weight, *disposable just like* anything that gets hauled off and dumped in a landfill."[30] Similarly, in *Wasted Lives*, Zygmunt Bauman writes that "to be declared redundant means to have been disposed of because of being disposable—just like the empty and non-refundable plastic bottle."[31] While this work has a great deal of affective power, this metaphoric usage is only a starting point for theorizing disposability.[32] This has become especially apparent in light of the sheer scale of the transnational flows of waste, pollution, and toxicity that characterize the capitalocene. Kathleen Millar points out that the language of human disposability is powerful but inadequate because it takes waste for granted and does not attend materially to economies, ecologies, and forms of living that take place after disposal.[33] With this black-boxing of waste as abjection, equating discards with bare life, Millar argues, the metaphor of disposability reaffirms a narrow liberal notion of human life and fails to recognize what Sylvia Wynter calls the multiple "genres of the human." This failure to engage substantially with the materiality and sociality of garbage undermines the intended critique of human disposability by compounding a racializing view of waste worlds' inhabitants as outside the human.[34]

Rather than discard disposability, the ethnography that follows attempts to move from a metaphor of disposability as an existential condition to a material and discursive analysis of the infrastructures that produce, sustain, and distribute disposability and the worlds it produces. Disposability is generative. The reason to linger on the nature of and infrastructures underpinning disposable commodities has been to draw out four central features of disposability—surplus, embodiment, displacement, and contestation—that I develop as analytics to understand waste worlds and their infrastructures. To understand the ways environmental injustice shapes worlds, produces vulnerability, and distributes injury, it is necessary to understand the material, infrastructural, and social processes of wasting. To consider people disposable *like* garbage sets up a parallel that takes disposability as self-evident and obscures the mechanisms through which people become disposable *with*, *through*, and *alongside* garbage. Analytically, the concept of disposability extends the focus beyond traditional economic domains of production, distribution, and consumption, expanding critical understandings of the social relations and cultural forms through which inequalities are made. Exploitation does not just occur in work sites; accumulation by dispossession does not only take place around extractive industries; and identity making does not only occur through shopping practices. Ethnographically, this means attending to the infrastructural, ecological, moral-material, ideological, and representational ways that disposability is produced and sustained, and attending to the ways that lives cast as disposable are, nonetheless, lived.

IV. INHABITING WASTE WORLDS

Slums have come to stand in as the definitive icons of African, if not all postcolonial, urban life and the ultimate spatial manifestation of global disposability. An estimated 85 percent of Kampala's residents live in the city's thirty-one slums.[35] Kampala's slums are located mainly in between the hills that shape the city's identity, in low-lying former wetlands once designated as wastelands by colonial surveyors. Because of the city's topography—and its incredibly complex political history of land tenure regimes—Kampala's slums are not a sprawling periphery but patches of poverty that crosscut the city, an unavoidable reality that inserts itself into the fabric of everyday life and infrastructure. Slums become waste worlds in a number of ways. As low-lying lands, they become reservoirs where multiple flows of uncollected wastes from elsewhere accumulate, mingling with local garbage that accrues in these underserviced areas. They are also the spaces where many waste workers reside and where the waste economy thrives. Precariously occupying valuable lands, they are also themselves cast as waste and displaced by speculative real estate investments and new infrastructure projects.[36] Because of these overlapping entanglements, my understanding of waste worlds emerges in conversation with the rich ethnographic literature that examines the economic and social relations of property and power that govern slums, the creative energies of residents who have built them, the everyday practices of inhabitation through which they become home, and the political struggles for alternative visions of development that counter precarity and displacement, as well as the marginalizing moral and technocratic discourses that pathologize them.[37] Slums represent a social, political, and environmental impasse that Kampala shares with many other African and postcolonial cities. They embody the paradoxes of economic growth paired with radical economic inequality in contexts of massive structural unemployment where wage labor, despite its absence, remains a definitive social relation that defines the imaginary of urban belonging. The emergent forms that cities are taking in places like Kampala challenge the foundations of theories of urbanization and citizenship premised on industrial employment and state-led planning processes and pose critical ethical questions of what to do with, how to think about, and how to relate to the huge numbers of people who have not been integrated into formal economies or traditional nationalist narratives.[38] The realities of the emergent urban world thus require a dramatic rethinking of urban and political theory.

As with alternate, and equally imprecise, terms like *informal community* or *low-income settlements*, the label *slum* circulates widely in Kampala among planners, policy makers, and residents of all classes, revealing the unstable and porous line separating emic and etic vocabularies. As is the case in the ethnographic contexts I describe, I use these terms somewhat interchangeably. While recognizing the pejorative implications of the term *slum*, I continue to use it for two reasons.[39] First, it names a way of seeing and problematizing urban space that many of the municipal workers I met share among themselves and with a global network of development practitioners. Second, as used by many residents of low-income neighborhoods, the term indexes the everyday indignities and forms of infrastructural violence that they endure and, through personal and collective means, seek to overcome. *Slum* thus names both a mode of governing urban space and a violent assemblage of socio-material conditions whose reality ought not to be erased through more polite managerial language.

Rather than exceptions to the urban norm, the slums of cities like Kampala characterize life for the majority of the world's urban population. Are these slums spaces of privation or of invention? Of abjection or of creativity? Of passive suffering or of agential reinvention of the urban form itself? Such questions have structured an overly polarized media debate that tends to oscillate between either/or positions while reinscribing the slum as a singular, ahistorical, and ultimately unknowable entity.[40] The response to normative models that can see only dysfunction and absence (the lack of quality housing and of jobs, sewerage, planning, etc.) has simply been to invert them through a form of celebratory functionalism through which anything making African lives viable is interpreted as entrepreneurial invention or resilience. The dystopian image of the slum has been rewritten by framing it as a laboratory resource for the project of global urban design.[41] The dominance of the slum as icon of African urban life has generated an Afro-optimist backlash seeking to highlight the positive, beautiful, cosmopolitan, and uplifting face of African cities in popular media. While there are important politics in diversifying the representations of Africa, this reaction has served to fetishize spaces of elite consumerism and violently occlude the reality of urban life for the majority of Africa's cities' residents, simply wishing them away so as not to embarrass the cosmopolitan elite.[42] Boosterish imaginaries of cities without slums are part of an "economy of appearances" that exemplifies how "Africa rising" narratives promoted by governments as a means of attracting foreign investment and conjuring new infrastructural projects into being.[43] This binarized discourse tells us little about the everyday struggles and practices through which African urbanites make and remake their cities, construct infrastructures, sustain social life, inhabit inequalities, and imagine better worlds.[44] Moreover, by erasing these routine realities, the oscillation between optimism and pessimism centers the Western gaze and replicates as cultural politics the slum-clearance characteristic of "world-class" city-making projects.[45]

Rather than frame a spectacle of poverty or succumb to elitist mandatory optimism, I approach Kampala as an "ordinary city," that is, as a city with a particular history and phenomenology that emerges through the mundane practices, routines, and relations of its diverse inhabitants.[46] While *Waste Worlds* is an ethnography comprising descriptions and analyses of these ordinary and everyday practices in marginalized urban spaces, it is not an ethnography of poverty as such. Rather than identifying, locating, and mapping poverty onto discrete territories like slums, I deploy a relational approach to poverty and prosperity, following Ananya Roy, to identify the ways in which these are socially and spatially arranged, examine the effects of such territorialization, and question the techniques of government that produce them.[47] How, and to what effect, are spaces differentially labeled as clean or dirty, developed or backward, valuable or wasteful?

Just as settler colonies did for eighteenth- and nineteenth-century Europe, today the slum functions as a spatial fix for rural dispossession and the production of a surplus humanity that cannot be absorbed into the circuits of formal capitalism. As Tania Li observes, in the context of structural unemployment and jobless economic growth, "The promise that [agrarian] modernization would provide a pathway from country to city, and from farm to factory, has proven to be a mirage," a broken promise fueling rural-urban migration.[48] In Uganda, agrarian displacement is driven by large-scale plantation developments across the country focusing on monocrops of palm oil, tea, sugar, eucalyptus, and pine, as well as by land speculation in the oil fields region bordering Lake Albert and the rural areas surrounding Kampala.[49] Urbanized surplus populations also exceed the grasp of Eurocentric forms of anticapitalism. Faced with economic informality and structural unemployment, they do not congeal into the expected class categories that form the traditional basis for leftist political theorizing and organizing.[50]

Despite their liminal position in social theory and political ideology, it would be wrong to understand surplus populations as existing outside of capitalism. Informality, for example, is often a disguised form of highly exploitative precarious wage labor.[51] Moreover, recent development strategies like bottom-billion capitalism that advocate entrepreneurial-ism and microfinance, as well as innovations in telecoms and fintech, have made low-income urban residents central to new modes of accu-mulation in which debt, rent, and data—rather than the wage—are the primary relations of exploitation.[52] Far from excluded from or cast out of the economy, surplus populations marginalized from formal wage-based employment are subject to capitalist social relations of predatory inclu-sion, adverse incorporation, or urban involution.[53] A socio-material approach to disposability is a useful analytic for understanding this rela-tion of superfluity—of materials and of populations—because it focuses attention not on dynamics of inclusion and exclusion but on violent and dispossessing forms of economic integration.[54] If disposable goods are a product of what Raj Patel and Jason Moore call cheap energy and cheap nature within capitalism, then disposable people, or cheap lives, are pro-ductive of cheap care and cheap work.[55] Disposability, then, cannot be understood purely as a form of abandonment but rather as a mode of inclusion, investment, and accumulation that produces waste.[56]

A key reason that urban youth are ideologically constructed as a sur-plus population is that African nationalisms in general, and Uganda's National Resistance Movement (NRM) in particular, have privileged the rural over the urban poor. As I discuss in more detail in part 3, at the core of NRM ideology is a vision of national development centered on the shift from subsistence to commercial agriculture through the educa-tion and infrastructural provisioning of a progressive peasantry. President Museveni presents his own vast cattle ranch as the model Uganda's peas-ants should emulate. Cautioning against land fragmentation, his vision of national development urges the population to commercialize farming and to manage risks through four-acre farms, divided between cash crops, dairy, and subsistence foods.[57] The NRM, in power since 1986, justified its ten-year ban on political parties based on the argument that because Uganda has only one class, the peasantry, political competition would be divisive and contrary to the aims of achieving national unity.[58] This devel-opment formula has no place for the urban poor, who—cast as matter out of place—are urged to return to their villages, the authentic heart of the nation. President Museveni, the erstwhile inspector general of police Kale Kayihura, and the city's journalists commonly refer dismissively to Kampala's youth as lumpens: disorderly and disreputable, undisciplined, and unemployed, an obstacle to development.

Marx was famously suspicious of the lumpen classes, dismissing them as "the garbage of society,"[59] a dangerous, manipulable, and reactionary obstacle to working-class solidarity.[60] Fanon, however, saw this population as the key to anticolonial revolution. Neither trapped in feudal structures like the peasants nor granted the privileges that turn the urban working class into a nationalist bourgeoisie with "everything to lose," Fanon regarded this residual class as the spearhead of urban insurrection: "That horde of starving men, uprooted from their tribe and from their clan, constitutes one of the most spontaneous and the most radically revolutionary forces of a colonized people."[61] For Fanon, the lumpen proletariat were not just "the sign of the irrevocable decay, the gangrene ever present at the heart of colonial domination," they are "like a horde of rats; you may kick them and throw stones at them, but despite your efforts they'll go on gnawing at the roots of the tree."[62] This dehumanization of the colonized, Fanon argued, is an effect of the colonial production of a racialized, disposable, urban surplus population dislodged from tradition and yearning for the birth of a new world. While I draw on Fanon to understand the politics of disposability, my ethnographic analysis does not set out to identify the true bearers of revolutionary history (nor did it inadvertently find them). Instead, I examine the forces, feelings, and routine institutions that channel popular political aspirations to describe the ways in which mundane popular infrastructures embody inequality, materialize dissent, and constitute a nonrevolutionary politics of space making and claiming. Understanding these contemporary forms of surplus and inequality in terms of disposability means tracking how people, populations, and practices are enrolled in processes of accumulation and urban development while simultaneously being expelled from the futures they construct.

Focusing on the material processes of discard is useful to understand the contemporary urban governance of poverty because displacement happens alongside disposal. As I take up in detail in part 2, both are equally predicated on the fantasy of Away—the idea that social and political problems have a straightforward geographic solution, throwing waste away and moving it out of sight.[63] Away is the spatial fix through which the contradictions of economic growth and radical inequality—mass urbanization without mass employment—are managed. The concept of sinks, drawn from environmental studies, offers one way of locating and understanding the spatial dynamics of Away. Sinks are sites—both places and bodies—that absorb and contain discarded surplus. They are reservoirs where pollutants, nutrients, or other types of compounds (carbon, for example) are stored.[64] Sinks enable the fantasy of endless growth by spatially distributing externalities in places where they aren't seen to matter.[65] Landfills are one example of a planned sink, the designated burial grounds of the refuse of modern lives. But sinks are rarely stable and are highly prone to spills, as the endless and endlessly laborious work of containing the waste from Japan's 2011 Fukushima nuclear disaster makes clear.[66] Oceans have become sinks; the famous Pacific garbage patch being only the most charismatic example of their current role as repositories of disposable plastics. Animal bodies become sinks, like the Puget Sound salmon found with flesh full of cocaine and antidepressants routed there through the city's plumbing infrastructure.[67] Humans also become sinks, bearers of toxic body burdens and evidence that the petrochemical environment of the Anthropocene entails a new era in anthropology as well as geology.[68]

Slums also become sinks. They are injurious hybrids, violent natural-cultural assemblages. Kampala's slums lie low between hills where debris flows downhill, literally washed down the class structure by torrential rains. Garbage accrues. Drainage channels clog. Streets flood. Fatal and endemic hepatitis B spreads. Cholera and typhoid break out. These are the diseases of poverty—preventable and treatable, but lethal. They create a heavier burden for the poor and actively contribute to the reproduction of poverty. As sinks, slums are sites of slow violence, "calamities that patiently dispense their devastation while remaining outside our flickering attention spans."[69] Garbage perfectly encapsulates slow, infrastructural violence: it is banal, impersonal, embodied, and quotidian.

In this light, shouldn't any effort to clean the city be applauded? Why develop an anthropological critique of the politics of cleanliness, when cleanliness is such a scarce and valuable good? Precisely because cleaning is a powerful form of world making, it is necessary to inquire into what worlds cleaners aspire to. *How* waste is posed as a problem matters.

Who is empowered to define and resolve that problem makes a difference. *Where* responsibility for the problem is located dramatically defines the contours of urban belonging. The argument that unfolds through this ethnography is that contemporary cleaning techniques produce disposability. While some philosophers argue that the Anthropocene requires new moral relations to waste and new ethical frames for inhabiting a filthy world, I am not interested in describing waste worlds as a desirable future or their residents as a moral avant-garde in living with damaged surroundings.[70] Waste worlds embody violent distributions of health and injury, and to pretend otherwise would be disrespectful to the people I came to know who make their homes there. This book does not make an argument against cleaning as such, nor does it offer a romanticized view of the politics of transgression. Rather, *Waste Worlds* is a critique of cleaning as a technology of displacement and of forms of urban development that reproduce disposability in the name of cleanliness. Because disposability is not simply about exclusion or abandonment, it is critical to understand how it is produced through forms of inclusion, investment, and care, like development, cleaning, and urban transformation.

Be it through the criminalization of waste work, the racializing rubrics of developmental respectability, or the institutionalized hierarchies entrenched in paradigms of inclusive growth, participatory planning, and community-based development, the imperative to clean up cities too often results in anti-poor politics, implicitly predicated on wishing away vast swaths of the population. The power of development discourse resides in its ability to attach popular desires for the material improvement of the conditions of life in a radically unequal world to a set of global institutions that systematically removes decision-making power from the most disadvantaged people. This ethnography is an effort to describe that hegemonic attachment and its effects.

V. INFRASTRUCTURAL VIOLENCE

November 16, 2007. President Museveni is on his way to Entebbe Airport to fly to South Africa. His trip is interrupted just six kilometers south of the state house when his convoy encounters heavy flooding on Entebbe

Road at Namasuba. Stuck on the road with other commuters and residents of the low-lying, low-income, roadside settlements, the president inspects his surroundings and gives an impromptu speech. "Look at these *buveera* (plastic bags) and bottles. They are the root cause of all this flooding because they have blocked the drainage system. We shall deal with it."[71] Four people died in that day's floods. Three drowned in Zzana, a few kilometers south of President Museveni's delay. One was electrocuted when flood waters touched a live wire in Bwaise, another flood-prone impoverished settlement north of town. Officials blame settlers themselves for living in harm's way. They argue that no one should have built houses in wetlands to begin with, as doing so makes homes prone to flood, or simply diverts water to exacerbate flooding elsewhere.[72] Amid the flooding, one woman stops scooping water from her submerged home to greet a foreign reporter: "Welcome to Africa! . . . You like the way we live here?"[73]

If this flood illustrates the multiple infrastructural entanglements that constitute disposability, then the responses to it exemplify the convoluted moral-material infrastructures of responsibility through which accounts of disposability are rendered. Plastic bottles and bags are held responsible for blocking drains. As litter, they evidence the irresponsible disposal habits of consumers. Additionally, the urban poor are blamed for living in low-lying areas where their homes both flood and cause flooding. Left out of these accounts are Kampala's chronic housing shortage that makes wetlands appealing options for home building, the proliferation of manufacturing and industrial operations in wetlands sanctioned by the national government or built by members of the president's family, a piped water system unable to provide clean water to the majority of the city's residents such that markets have become the primary means of distributing water in the city, disposable plastic bottles and bags taking the place of pipes as urban infrastructure, and municipal waste collection rates estimated to be 30 percent, which explains the amount of trash in the city's drains.[74] The question is no longer *if* infrastructure has politics, as Langdon Winner asked forty years ago, but *how*.[75]

This question has animated a groundswell in studies of infrastructure in anthropology and beyond. While political theory has often relied on infrastructural metaphors to theorize the relationship between economics, politics, and culture, one of the central interventions of the "infrastructure turn" has been to pay attention to the materiality, governance, and political economy of these systems in order to theorize infrastructure itself as a means of organizing social difference, distributing power, remaking environments, and constituting states.[76] This research teaches us to notice how the lethality of Kampala's floods emerges from the intersection of multiple infrastructures—roads, housing, water, electricity, drainage, and waste management—and how these combine to differentially distribute injury across the population, materializing the city's radical inequalities and putting some bodies in harm's way. Another critical aspect of the infrastructure turn has been to theorize infrastructures' poetics: the meanings, narratives, moralities, and affects that accrue around them and coproduce socio-technical systems, organizing and attaching social formations to them.[77] Building on this work, my contribution in this book is to explore both the material organization of waste infrastructures and the discourses of responsibility and blame that surround them to track how, together, these produce disposability and moralize it in ways that exacerbate displacement. Accounts of flooding that emphasize highly visible proximate causes (plastic waste, flood-prone homes) and those imagined to be responsible for them, in this case, serve to sanction the demolition of slums and the eviction of the supposedly polluting populations who inhabit them. These technocratic framings of the problem of waste and cleanliness are themselves part of the moral infrastructure of disposability insofar as they subtend the social and environmental processes of displacement.

Infrastructural violence refers to the interconnected ways that infrastructures participate in slow and structural, as well as more traditionally conceived, forms of violence.[78] If disposability is a socio-material condition, infrastructural violence is its central mechanism, the *how* of disposal. First, infrastructures can be designed to inflict violence; think of border walls, prisons, and even the small-scale hostile architecture implanted in everyday urban space.[79] Infrastructures are also targets of violence, as in the case of bombings targeting train stations in India or the destruction of the urban fabric in Palestine.[80] Second, infrastructures can be the means by which structural violence takes place, be it through forms of explicit exclusion—as in the famous example of Robert Moses's bridges designed to keep public buses and thus Black and low-income New Yorkers from accessing public recreational facilities—the planning of hazardous and unwelcome infrastructures like incinerators, or the uneven provision of services like water, electricity, and sewerage.[81] Third, infrastructural violence refers to the forms of displacement that occur when informal infrastructures are criminalized and the livelihoods they sustain are eradicated.[82] As with the concepts of structural and slow violence, infrastructural violence emphasizes the ways in which these forms of violence rarely constitute recognized events, instead remaining as naturalized background conditions of everyday life.[83] The concept offers an added emphasis on the materiality of these processes and their complex imbrication in forms of planning, regulation, privatization, and normative narratives of urban life and urban futures.

Throughout this body of research, infrastructure is double-sided, serving as both an object and an analytic. As object, infrastructure is a social fact: roads, pipelines, dams, servers, satellites, and all the other socio-technical systems that constitute the grid that underpins modernity, as well as a political promise and social aspiration, a category of financial investments, and a techno-political apparatus of planning, funding, and management. As an analytic, infrastructure highlights what Brian Larkin calls the "peculiar ontology" of these objects that are "things and also the relation between things."[84] In this sense, infrastructure identifies this relational quality in other phenomena like spatial arrangements, kinship, law, affect, labor, morality, categorization systems, and more. Rather than a discrete category of objects, then, any "thing" can be an infrastructure when it manifests this relational quality of being a means by which some other object, condition, or process is achieved.[85] The infrastructure of disposability, in this sense, refers not simply to a complex socio-technical system of waste management that moves garbage through Kampala in order to keep the city clean but moreover to the heterogeneous assemblage— of waste, technologies, discourses, moral logics, forms of work, political rationalities—that produces, legitimizes, and distributes the social and environmental condition of disposability.

Infrastructure occupies a crucial economic and ideological position in structurally adjusted Africa. New projects disclose the shifting boundaries between state, government, and market while infrastructurally enacting, institutionalizing, and materializing Africa's place in the world.[86] As in the immediate post-independence period, infrastructure projects continue to play a symbolic role as national flagships, invested with both the developmental aspirations of the population and debts that will endure over generations.[87] However, while the nationalist infrastructures of the 1950s and 1960s were built under the rubrics of modernization theory and were designed to foster the emergence of import substitution economies, today's roads, pipelines, ports, dams, and export processing zones reveal a more directly extractive set of growth strategies.[88] These infrastructures—being constructed at vast expense through new partnerships between African governments, international financial institutions, transnational corporations, foreign governments (increasingly the Chinese government)—are the material basis of global flows that link Africa's resources to external markets, with benefits largely bypassing the continent's population even as they reshape local lives.[89] Likewise, since neoliberal rationalities of rule have become hegemonic, vital municipal services like electricity, water, and waste management across the continent have been contracted out through public-private partnerships.[90] Urban infrastructures are thus not simply a key means of establishing the healthy business climate necessary to attract foreign direct investment, they have themselves become sites of investment to be reorganized as competitive markets.

Compared to the spectacular hydroelectric dams being built on Uganda's stretch of the Nile River, the complex network of refineries and pipes being planned in the oil region surrounding Lake Albert, or Kampala's logistically mind-boggling shipping-container forwarding yards that link the city's consumers to transoceanic shipping routes, municipal waste management is distinctly unglamorous, even in the field of infrastructure studies foundationally established as the study of "boring things."[91] It does, however, afford a less state-centric view of infrastructure. Because the barriers to entry are far lower in recycling than in projects like dam building, waste management is, by comparison, a hugely participatory, decentralized, heterogeneous, and everyday infrastructure.[92] Garbage is, in one way or another, a routine part of daily life for every resident of the city. Making waste infrastructure work requires configuring users to systems through "awareness raising" and "behavior change," techniques reliant upon highly moralized discourses of responsibility, respectability, and citizenship. Infrastructure, in this way, comes to take place in the minds of urban residents. As ways of governing waste also become ways of producing subjects, mentality is recast as infrastructure.[93]

Waste management is very expensive. A huge question facing municipal planners is how to pay for a service that represents around 10–25 percent of their annual budget.[94] Cost recovery has been one answer. Rather than subsidizing low-income areas with free garbage collection, various experiments have been made to charge for the service. None have succeeded. In contrast to other critical services like water and electricity, garbage infrastructure is about the disposal rather than acquisition of matter, a fact that informs the everyday practices and social relationships that emerge in waste worlds, and shapes contestations over citizenship and belonging. Waste is simply too easy to dispose of otherwise by burning, dumping, or burying. While these practices are illegal, they are easily accomplished under the cover of darkness, by children, or out in the open with the impunity of universality. Private waste companies do not bother marketing their services in the city's unprofitable poor neighborhoods, knowing that no one will be able to afford them. Despite the failure of cost recovery in the formal sector, small-scale waste collectors serve Kampala's poorest neighbors and earn a living doing so. Their services are irregular and unpredictable, and they often dump the wastes they collect in wetlands. They have, however, been able to develop waste collection infrastructure that removes waste from poor households using simple tools like wheelbarrows and modified bicycles. Charging USh100–500 (US$0.04–$0.20; *USh* denotes Uganda shillings) to remove a sack of rubbish, they have developed complex credit and payment schemes with regular clients that choreograph the flow of waste and the unsteady cash flows of the urban poor. For all its economists and social researchers, the World Bank has not been able to do this. State-centric infrastructure planning, both modernist and neoliberal, however, regards this accomplishment as a nuisance, a polluting and unlawful enterprise. Incorporating such systems into the official waste stream has been done in other cities in Africa, Asia, and Latin America.[95] Regardless, the KCCA criminalizes, arrests, fines, and impounds the equipment of informal waste collectors in the name of cleaning the city, insisting that waste entrepreneurs acquire prohibitively costly environmental impact assessments. Such forms of criminalization and displacement disavow the complex relations that exist between official and popular infrastructures, the former in many ways depending on the latter for its own success, as I outline in part 2. In this way, infrastructural investment can render other infrastructures disposable, defining the contours of urban belonging by undoing the diversity of city-making practices that constitute Kampala.

Despite the ways they are socially devalued and the symbolic pollution they accrue, waste worlds are far from being sites of pure abjection. On the contrary, garbage infrastructure is a generative location for the emergence and expression of competing regimes of value and urban belonging. Working with waste enables diverse Kampalans to make a wide range of claims to urban citizenship. These include salvagers' advocacy based on their recognition of the vital function they play in managing waste, as expressed succinctly in one interview by a young man who demanded to know "how can they [the KCCA and the assumed elite interests it represents] live without us, when without us they would choke on this rubbish!?" They include explicit discourses of good citizenship—elaborated in constitutive opposition to figures of unenvironmental, irresponsible, and polluting others—that shape volunteers' feelings at high-profile cleaning events. They include the struggling cosmopolitanism of impoverished community-based development organizations looking for ways to connect their small-scale projects to transnational streams of NGO resources in order to make better lives and livelihoods for themselves and their neighbors by making something of the waste that threatens their homes and health. They include the quiet encroachments of plastic-bottle collectors working the city streets at dawn so they can rent a room for the month, feed themselves and others, and imagine building a less precarious presence in the city. Precisely because they are so mundane, yet so contentious and contested, waste worlds are rich in claims, practices, and demands for belonging, citizenship, and justice.

Waste Worlds makes the argument that disposability is best understood not as an existential condition of social exclusion but as a form of inequitable social inclusion that combines investment and material abandonment, functional dependence and legal criminalization, care and blame. Disposability is an infrastructural condition. People and places become disposable not by being cast out of the city but by being held in a particular relation to it: collectively vital and necessary for the reproduction of urban life but individually expendable and exposed to the harms of waste worlds. Waste worlds are not abject voids but rather densely inhabited and hugely generative. What follows builds this argument in three parts. Part 1 explores waste as an object of governance in order to understand how waste management politics have been implicated in the constitution of political authority throughout Kampala's history and how, since 2011, new articulations of technocratic and authoritarian rule have been manifested in projects to clean the city. Part 2 explores waste as an object of labor to illustrate the quantity and heterogeneity of work that goes into producing waste infrastructures and the proliferation of economic and social projects that this work entails. Part 3 explores waste as an object of affective investment in order to understand the entanglement of care and abandonment in waste worlds and to track the ways that racial ideologies inform and emerge from Kampalans' encounters with waste and with each other. Throughout all three parts, I emphasize how material *surpluses* and governing ideologies about surplus populations shape the formation of waste worlds. I track how waste worlds are *embodied* through toxic exposures, everyday practices of cleanliness, and the routines of waste work. I detail how imaginaries of a clean city and the forms of development being promoted in Kampala *displace* waste worlds. I attend to the forms of *contestation* that these dynamics engender, as waste worlds' inhabitants make worlds for themselves and challenge the logic of disposability.

PART I   The Authority of Garbage

1 Accumulations of Authority

> Thus says the Lord God, "On the day that I cleanse you
> from all your sins I will also cause the cities to be inhabited,
> and the ruins will be rebuilt. The desolate land will be culti-
> vated instead of being desolation in the sight of everyone
> who passes by. Then they will say, 'This land that was
> deserted and desolate has become like the Garden of Eden;
> and the waste, desolate, and ruined cities are fortified and
> inhabited.' Then the nations that are left round about you
> shall know that I, the Lord, have rebuilt the ruined places,
> and replanted that which was desolate; I, the Lord, have
> spoken, and I will do it."
>
> —Ezekiel 36:33–36

In a 2014 interview reviewing her accomplishments as executive director of the KCCA, Jennifer Musisi cites Ezekiel 36:33–36 as a source of purpose and inspiration as she takes on the task of transforming Kampala into a clean, orderly, and functioning city.[1] Ezekiel relates an apocalyptic prophecy of divine judgment, devastation, and redemption, prophesying God's wrath: "Wherever you dwell your cities shall be waste and your high places ruined."[2] It tells that the people of Israel will be scattered into exile, punishment for the sins of idol worship, violating taboos, ritual transgressions, and spiritual uncleanliness that profane God's name. Establishing the Lord's sovereignty over a host of foreign nations, Ezekiel relays God's proclamation that He will "make the land desolation and a waste."[3] After suffering God's wrath, the people of Israel will receive His beneficence; Ezekiel prophesies a gathering of the scattered people on the lands of their ancestors where God will deliver them from uncleanliness, give them a new heart and a new spirit, and transform waste to wealth, turning the defiled desolation into a fertile land for a fertile, righteous, and resurrected people.

While it may seem exaggerated to compare this biblical narrative of the judgment and redemption through which God's sovereignty is proved and a people is reborn to the story of municipal transformation in contemporary Kampala, Musisi's reference to this passage attests to some of the moral narratives through which urban politics in Uganda are understood. The KCCA, controversially established in 2010 to restructure urban governance, is not just a technical project but an agent of moral transformation intent on turning a wasteland into a prosperous and Edenic city. Musisi observed that, prior to the KCCA, "Government and all Ugandans were constantly embarrassed about the state of the city."[4] So, as the first head of the KCCA, she set out to repair the image of the city and transform the nation's feelings about its capital.

The following chapters show that garbage served as a material, practical, and symbolic foundation for the KCCA's authority, signaling its mandate and its ambitions as well as the infrastructural terrain upon which its legitimacy would be established. This is surprising because waste management provides few occasions for the kinds of elaborate white-elephant projects through which a new government might seek to make its mark on a city. Nonetheless, of all the issues facing the city—potholed roads and traffic congestion, wetland encroachment, flooding, housing shortages, health care, air quality, food security, and the overriding concern for the urban population and unemployment—in its first year of operation, the KCCA made solid waste management its first priority. Routine maintenance emerged as the critical field in which the KCCA's highly contested governmental restructuring was legitimized and a new municipal government authorized. What does the fact that a newly established municipal government sought to found its authority on garbage reveal about the landscape of government and the constitution of political authority in Kampala? How does routine maintenance constitute, consolidate, and reproduce municipal power and urban space? Garbage is a messy foundation for political authority, one that opens a lively terrain of popular contestation. Waste management, one of the most unspectacular and mundane aspects of urban governance, offers a privileged view into the establishment of the

KCCA, how it legitimized its authority in the face of protest, and how it has pursued its project of antipolitical urban renewal. Waste thus emerges as central to world making. As an object of governance, it becomes a generative substance through which social and political relations are made and remade.

This chapter situates the contemporary politics of cleanliness in relation to changing forms of political authority since the early twentieth century. A chronic crisis of rule, authority, infrastructural disorder, and uncontrolled development (the condition sometimes labeled with the Luganda term *kavuyo*) in Kampala were the historically situated problems that the KCCA was established to resolve. Putting this crisis of rule into a longer historical view illustrates the ways contemporary political struggles continue a pattern in which waste infrastructures and the moral economies of cleanliness operate as both a source of legitimacy and a terrain of political contestation. It reveals how routine maintenance work like waste management is simultaneously shaped by and productive of the ongoing multiplication of political authority in Kampala. Infrastructural maintenance, I argue, is also the maintenance of the forms of power and political authority that infrastructure sustains.

To call Kampala the seat of government in Uganda would be a radical understatement. The history of the city is a history of the accumulation and sedimentation of multiple forms of authority. Kampala's topography of rolling hills interspersed with swampy lowlands suggests an alternative to theories of sovereign power inspired by the singular peak of Hobbes's Leviathan. Rather than a lone Hobbesian sovereign authorized by the mutual covenant of a terrorized people, political authority in the city is multiple and heterogeneous. While most of the ethnographic research that informs this book is concentrated on the city's wetlands, drains, and dumps, this chapter heads for the hilltops to describe the proliferation of authorities that shape the political terrain upon and against which visions of a clean city take shape.

In the century prior to the establishment of British over-rule with the 1900 Buganda Agreement, the hills that make up contemporary Kampala were the political center of the Buganda Kingdom. Each new kabaka (king) founded a new capital on a hilltop, establishing a palace, administrative center, and court. Upon the king's death, his palace became a royal tomb,

a site of spiritual authority tended by mediums, while the successor established a new capital.[5] Recognizing the capital's strategic location, and
perceiving some commensurability between the Buganda Kingdom and
Victorian Britain—their hierarchical centralized political structure and culture, the value placed on progress, and an emphasis on domestic virtues
like cleanliness—British officials saw the Buganda Kingdom as an attrac-
tive partner in the project of indirect colonial rule.[6] The British, in turn,
consolidated the kabaka's authority at a time when the kingdom was
wracked by a territorial war with the neighboring Bunyoro Kingdom and
an internal religious conflict pitting adherents of the Catholic and Anglican
churches against each other.[7] Missionization preceded colonization and, by
virtue of the conflicts it generated, played a role in fragmenting and desta-
bilizing the kingdom, which facilitated the establishment of colonial rule.[8]
Under the British, the Buganda capital ceased moving and was consolidated at Mengo, the Lukiiko (the Buganda Parliament comprised of chiefs
appointed by the kabaka) was established nearby on Rubaga Hill, and the
tomb of Kabaka Mutesa I at Kasubi became designated as the Buganda
Royal Tombs.[9] Although kingdoms were abolished in 1966 under President
Milton Obote, who saw the kabaka as a threat to his own authority, in 1993
they were reinstated as purely "cultural institutions" by President Museveni in a bid to consolidate his own authority and popularity in Buganda.[10]

In the 1890s, prior to the consolidation of British rule, Frederick
Lugard established a fort on a hilltop neighboring the palace of Kabaka
Mwanga II at Mengo, under the authority of the Imperial British East
Africa Company. This fort—Camp Impala, named after the antelope that
made the area a rich hunting ground—became the nucleus of the colonial city of Kampala that extended from what is now known as Old Kampala to Nakasero and Kololo Hills and stood apart from the Kibuga, the
native town centered around Mengo that remained under the authority
of the kabaka.[11] Kampala was a segregated city, inhabited by British merchants and colonial officers as well as Indian traders and laborers. The
neighboring native town was home to a large African migrant population from around East Africa, while the majority of Baganda who worked
in Kampala commuted from nearby villages where they resided.[12] The
1903 Township Ordinance established a planning framework based on a
system of cantonments imported from colonial India. Planning schemes

were drawn up based on the recommendations of staunch segregationist Professor W. J. Simpson in 1912, and again in 1919, to establish cordons sanitaires between European and Asian areas but did not incorporate any African settlements.[13] Not until 1947, when avant-garde German architect Ernst May was hired to create an ultimately unimplemented plan for the rapidly expanding post–World War II city were African workers incorporated into municipal planning. May proposed a garden city model including discrete African quarters with the ambitious modernist goal of carefully evolving Africans through holistic special, social, and cultural planning.[14] In the early 1950s the final British plan for Kampala was proposed, and it responded to growing demand for African housing and urban rights by incorporating the Nakawa-Naguru area as a planned African residential area. For African workers, inclusion still involved segregation. As Ann Stoler has argued, however, it is important not to overstate the fact of segregation and portray an entirely divided society, ignoring the imperial intimacies and circulations across boundaries constitutive of colonial racial orders.[15] On one hand, colonial rule created elaborate campaigns to intervene in and remake African intimacies, including efforts to abolish nonmonogamy, promote fertility, and control disease.[16] On the other hand, British colonists' domesticities relied upon African labor in the most intimate spaces of European homes, as the cleanliness of white spaces required the circulation of African workers across the urban borders that racially demarcated space.[17]

In Kampala, as throughout the British Empire, waste and cleanliness were central to the construction and maintenance of racial difference, both between colonizer and colonized and between differentially racialized members of the colonized population.[18] In his study of the Buganda capital late in the colonial period, anthropologist Peter Gutkind observes that early explorers, missionaries, and colonial administrators in fact offer contradictory, but consistently racializing, accounts of sanitary conditions and the importance placed on cleanliness by the kabaka. These range from statements that label Baganda as a rare "Negro race . . . who attempt anything like sanitary measures to keep their surrounding areas from filth" by Special Commissioner Sir Harry Johnston and missionary ethnographer John Roscoe's report that the kabaka "made the neglect of certain sanitary conditions in the capital an offense punishable by death" to medical officer Dr. Ansorge's depiction of the "abominable filth in the native capital" with "filth and stench everywhere."[19] Responding to this filth, cleanliness, hygiene, and sanitation became a struggle as the Kampala Municipal Council sought to extend sanitary authority over the Kibuga, meeting both cooperation and resistance from the kabaka and his prime minister.

Sanitary measures in the 1920s and '30s included using forced labor to clear drains, burning huts, and hunting rats, primarily with the goal of combating plague and malaria.[20] Although the outcome of these efforts was often appreciated, following the orders of colonial medical officers was seen as compromising the king's sovereignty over the Kibuga. Carrying out these orders also made Baganda authorities unpopular among their followers, upon whose labor they relied and who were being evicted from unsanitary huts.[21] In his 1930 masterplan, A. E. Mirams, a town planner and fellow of the Royal Sanitary Institute with prior experience in colonial Bombay, proposed constructing a refuse incinerator outside Kampala in an area that appeared blank on his maps but was an inhabited part of the Kibuga. Mirams noted simply that the "site is on native land but can be acquired."[22] The disposal facility would be "screened from the public gaze, and so sited as to create as little annoyance to the public as possible."[23] The white public would thus be spared the annoyance of encountering its own waste and the infrastructures of disposability by expelling them from Kampala into the African city.

Racialized dualism marked both the development of Kampala and the ways in which the city's social problems were understood. Diagnoses of urban ills disavowed relationality between colonial and African settlements and, as Achille Mbembe writes, "circumscribe[d] the phenomenon of poverty within racially associated enclaves."[24] Writing in 1961 at the eve of Uganda's independence, for instance, political scientist David Apter observed that Kampala is "prepossessing" and "exudes prosperity. . . . It has no 'African Quarter'" although "there are slums *outside*, such as Kisenyi and Mulago, sometimes quite fearful ones, but by and large African immigrants live there."[25] This distinction between Kampala and its outside, the Kibuga, was a sustaining effect of a political and infrastructural dualism that gave rise to a clean Kampala on the one hand and "squalid slums in a royal realm" on the other.[26] As in other East African cities, cleanliness was thus established as a critical terrain of political contestation in colonial Kampala.[27] It marked the contours of urban belonging, materialized racialized theories of who counted as a proper urban citizen, and legitimized or delegitimized particular forms of rule.

When the kabaka was exiled and the kingdom abolished by President Obote in 1966, Kampala and the Kibuga were integrated under the authority of the Kampala City Council. A new integrated plan for the city was prepared in 1972 but, amid the chaos of the Amin Regime (1971–1979), was never implemented.[28] Kampala decayed during these years and over the course of the ensuing Obote II regime (1979–1986), marked by civil war and authoritarian violence directed at the population. The Asian expulsion and economic war of 1972 led to the collapse of Uganda's industrial and manufacturing sectors. With the loss of jobs, stable incomes, high tariffs on imports, and state control of primary exports, a black-market economy—dubbed the "magendo economy"—flourished in these years, and informal means of securing an income were normalized across the class spectrum.[29]

Since Yoweri Museveni and the National Resistance Army (now the ruling party, the National Resistance Movement, NRM) took power in 1986, established peace in southern Uganda, rebuilt the national state, and restructured the national economy through IMF-authored structural adjustment policies, the city has experienced a boom in population, commerce, and international nongovernmental activity.[30] In 1993, prior to establishment of the nation's new constitution, a ten-year urban plan was commissioned by the Ministry of Lands, Housing and Urban Development.[31] By far the most comprehensive plan at the time, it cataloged the city's structural, environmental, sanitary, legal, medical, traffic, gender, and education conditions and proposed extensive measures to improve living conditions and institutional capacity in the city. This plan remained largely unimplemented as the municipal government and international donors responded in an ad hoc, project-based way to urbanization through the 1990s and the first decade of the 2000s, giving rise to the planning crisis that the KCCA was established to resolve.[32] From 1989 to 2010 the city's built-up area expanded over fivefold as new residents built precarious, unplanned settlements in wetlands and on steep hillsides, as well as spreading along major roads to incorporate nearby townships.[33] In 2007, UN Habitat estimated that the slums constituting 21 percent of the city's area housed 39 percent of its population. At current rates, it is projected that Kampala will sprawl to one thousand square kilometers by 2030—compared to seventy-one square kilometers in 1989—with "inhuman conditions for the majority of the urban population."[34] Characterized as a period of laissez-faire urbanism, the population and economic boom of the 1990s was accompanied by discrete donor-driven urban development projects rather than systematic planning, leading to the proliferation of congested informal settlements and outbreaks of cholera, as well as the consolidation of urban agriculture and the further expansion of the informal sector.[35] Manufacturing industries have reemerged in Kampala's designated industrial areas as well as in "free" special economic zones along the Kampala-Jinja highway, designed to attract foreign direct investment by providing necessary infrastructures and relaxing environmental and labor protections.[36]

The remaking of the Ugandan state under President Museveni has taken place alongside, or more accurately *through*, the growing presence of international financial institutions (IFIs) like the World Bank and the International Finance Corporation, as well as transnational and local NGOs. Contemporary Kampala is densely occupied by NGO and IFI offices, the city's major hotels thrive on their conferences and workshops, and their well-compensated "expatriate" staff are a major driver of real estate speculation and construction in the city fueled by Ugandan investors and landowners eager to earn rents payable in foreign currency.[37] Studying the topography of AIDS treatment in Uganda, the medical anthropologist Susan Reynolds White and colleagues argue that the "Ugandan landscape [of care] can best be described as projectified," citing the endless "list of acronyms for programs" funded by myriad international donors and coordinated by multiple national umbrella organizations with ties to the state.[38] This donor dependence, they point out, "became a strategy to strengthen the state," insofar as it accumulated legitimacy for the nascent Museveni/NRM regime in the context of the AIDS epidemic in the wake of the guerilla war.[39] While this "projectified" landscape, with its associated relations of clientship that configure belonging through the exchange of material support for behavioral discipline and personal narratives, is most dramatically illustrated in AIDS care, it also describes urban governance.

In addition to housing international NGO headquarters with a national remit, Kampala is home to NGOs engaged in a wide range of issues in the city itself, focusing on sanitation, education, public health, slum improvement, human rights, HIV/AIDS, housing, and entrepreneurialism in discrete local development projects. This projectified form of governance is not continuous, systematic, or evenly distributed across the city but patchy, concentrated, and punctuated. Government in the city (and nationally) is thus distributed across these networks of rule by NGOs that destabilize analytic distinctions between state and nonstate actors, local and global, sovereignty and government. NGO projects articulate with the multiple tiers of the municipality itself. Kampala is divided into five divisions, each with its own elected mayor and appointed town clerk, subdivided into parishes and local councils (sometimes referred to as villages), with each scale having its own elected officials. In addition, nine members of Parliament (MPs) represent the city. Parliament itself, along with nearly all the national ministries, is located in the city where it literally casts a shadow over neighboring City Hall.

Spiritual, medical, and scientific authority also emanates from Kampala's hilltops: Catholic St. Mary's Cathedral atop Rubaga Hill, the Church of Uganda's St. Paul Cathedral atop Namirembe Hill, and the Uganda National Mosque, built with funds donated by the late Libyan president Muammar Gaddafi on the site of Fort Lugard atop Old Kampala Hill. In addition to these hilltop centers, evangelical churches have proliferated around Kampala since the mid-1980s. Ranging from grandiose megachurches to the myriad upstart corrugated-iron churches of the city's poorest congregations, they have radically pluralized the sites and sources of spiritual authority in the city. Medical authority, historically entangled with spiritual authority and missionization, equally occupies high ground in the city. Founded by the Church Missionary Society in 1897, Kampala's first hospital, Mengo Hospital, sits just below Namirembe Cathedral, while Nsambya Hospital, established by the Little Sisters of St. Francis in 1903, currently caps Nsambya Hill. Mulago National Referral Hospital, a center of healing, research, training, and public health in Uganda, sprawls over Mulago Hill. Mulago's neighbor, Makerere Hill, gives its name to Makerere University, the pinnacle of the country's educational system and a bastion of scientific expertise and intellectual authority.

While it is difficult to overstate the importance of executive power in Museveni's Uganda, or the president's centrality in popular conceptions of power, politics, and the possibility of transformation, the politics of maintenance and cleanliness reveal a far more diffused and multifaceted landscape of political authority.[40] This more heterogeneous picture of authority offers a rich framework for understanding the actual operations of government and the foundations of authority in the city. It points to the diversity of actors doing maintenance and producing space in the city; the instability of the categories of state, government, and civil society; and the kinds of authority mobilized and produced through waste. The KCCA was born in the midst of this proliferation of regulatory, political, and moral authority. Founded by an act of Parliament with the mandate to transform the city, the KCCA's ambition was to centralize technical expertise and planning authority, becoming a relay point for development in the city. The KCCA sought to become a meta-authority, a clearing house for managing the projectified landscape of government, rationalizing the unevenly sedimented layers of political authority, and coordinating the maintenance of the city. Distributed across the urban landscape; produced in commercial, residential, and industrial areas; managed by public, private, and informal actors; and problematic for a range of health and aesthetic reasons, garbage proved to be an ideal capillary conduit through which the new authority could take shape. Chapter 2 describes the political problem that Kampala posed for the ruling NRM party and the way in which the KCCA was founded and authorized as the solution to a particular problem defined in terms of administrative chaos and developmental failure: a city in need of repair.

2 Tear Gas and Trash Trucks

Through the windshield of a brand-new trash truck ferrying waste from a neighborhood in southern Kampala to the landfill north of town, Musa Lubinga—a KCCA worker I was shadowing for the day as he supervised waste collection in his division—and I saw what looked like an occupying army installing itself in the center of a roundabout in the middle of busy Kampala traffic. Dozens of police and other security forces clad in blue and green uniforms milled around on the recently trimmed grass. Others sat, weapons in their laps, on benches mounted in the beds of pickup trucks, seemingly ready to zoom out into traffic at any moment. Behind them, an armored vehicle the size of the trash truck was parked with a water cannon protruding from its roof pointed casually at the circulating cars. "Is there a rally or something today?" I asked Musa, struck by the extent of the security buildup on an otherwise normal day in the city. "No, I don't think so. They're always doing this now," Musa replied, pointing out that he saw these vehicles parked around the city's major roundabouts most days he was driving. Indeed, while the scale of the security forces we observed that day in March 2013 was unusual, the presence of highly armed military and police forces in the middle of Kampala had become a routine sight. A conspicuous assertion of the state's repressive capacities, these displays were evidence of the rising number of police employed in the city and the military hardware they were amassing.[1]

These two vehicles, the armored water cannon and the trash truck, are emblematic of urban politics in Kampala, embodying the dual imperatives of securitization and service delivery that underpin urban governance in Museveni's Uganda. This chapter examines this confluence of militarism and maintenance, tracking how waste management emerged alongside militarized policing as a dominant solution to Kampala's disorder. How did maintenance become such a central terrain of political struggle, and what impact did the politicization of maintenance have on the forms of cleaning that emerged under the KCCA? While the state's repressive apparatus and its spectacles of violence putting down urban protest movements have received more media attention and political analysis, the mundane work of municipal maintenance was equally central to the formation of a new mode of urban government in the face of increasing urban unrest leading up to and in the wake of the walk-to-work protests that shook the city and the NRM regime in 2011.

President Yoweri Museveni's regime has been characterized as a hybrid of democratic and authoritarian, technocratic and militaristic modes of rule.[2] Museveni's unique political genius has been in balancing internal and international politics, materially and discursively gatekeeping. The patronage networks of an expanding state and security apparatus, as well as the rewards given to commercial elites through the privatization of state assets, have allowed him to co-opt and preempt both political dissent and armed resistance by bringing potential enemies into the NRM fold through positions in the proliferation of newly created districts or as commanders of new military and police forces.[3] His position as a key US geopolitical ally in the war on terror, enhanced since committing Ugandan troops to the African Union Mission in Somalia in 2007, ensures a steady stream of military aid as well as development funds, while his fealty to IMF structural adjustment plans famously earned him a reputation as a member of the so-called "new generation of African leaders," bringing democracy and development to the continent after the Cold War.[4] Until 2016, when a news bulletin summarized international skepticism in declaring that the regime was "not even faking it anymore," regular elections and the appearance of democratic institutions had been vital in maintaining Museveni's international standing.[5] Elections provided a veneer of democracy that has ensured the continued flow of international development funds and projects into the national budget, further consolidating Museveni's ability to fund elaborate patronage networks.[6]

Gatekeeper states, Frederick Cooper writes, inherit the structural position of colonial governments, "strong at the nodal points where local society meets external economy, dependent on manipulating revenues and patronage possibilities deriving from that point, including foreign aid and commercial deals."[7] As the capital city and economic center of the country, Kampala occupies this nodal point, and so controlling urban space has become particularly important for maintaining the NRM regime. Despite the NRM's grip on the national government, Kampala has been represented and ruled by politicians from the opposition Democratic Party since 1998, the city's first multiparty mayoral election of the NRM era. While winning national elections buttresses Museveni's legitimacy—even when their credibility is contested—losing them in Kampala complicates the NRM's ability to control the city. The 2010 KCCA act aimed to resolve this situation by rendering the capital city's highest elected official, the lord mayor, a powerless ceremonial counterpart to a presidentially appointed executive director. The act added the technocratic authority to maintain the city to the NRM's control over security forces, institutionalizing developmental authoritarianism at the urban level. This assemblage is crystallized in the city's small parks and garden roundabouts, which have been restored and beautified while remaining off limits to the public and offering strategic locations for military and police riot-control vehicles. The infrastructural power to clean and maintain the city, particularly to properly dispose of garbage, played a vital role in building the legitimacy of this new arrangement.[8] By taking on the project of urban maintenance and repair, the KCCA would repair municipal authority itself and bolster the technocratic credentials of the ruling party.[9]

How did Kampala come into disrepair? As Julie Chu argues in her work on the politics of urban renewal in China, disrepair can occur as part of political strategies aimed at securing and legitimizing urban transformation, making it seem necessary and uncontroversial.[10] Likewise, in Kampala disrepair is not natural or inevitable. It is a historical effect of efforts by the national ruling party to discredit the opposition and of decades of donor-driven policy guided by the imagined need to counter "urban bias" through structural adjustment plans that dismantled municipal budgets.[11] As in many other urban African contexts, opposition parties that have had no success winning significant representation in Parliament, let alone control over the executive branch, have found great success in cities by deploying populist rhetoric.[12] Opposition mayors' failures to develop the city once in power, they argue, has to do with the ruling party's unwillingness to provide sufficient budgetary support to enact reforms or carry out their policies in any meaningful manner.[13] Urban governance has been characterized by open confrontation between City Hall and the State House over decision-making and budgetary control. Supporters of the president and the National Resistance Movement accuse Kampala's elected representatives of being corrupt and incompetent rabble-rousing populists who appeal to and manipulate the urban masses to advance their own political careers but have no real vision for the city. Supporters of the opposition accuse the president of deliberately sabotaging urban governance by withholding budgetary support and tacitly sanctioning disorderly development (through unplanned land sales, allowing well-connected developers to ignore municipal and environmental regulations, and playing populist himself when convenient) in order to prevent the opposition from gaining the political capital and legitimacy they would accrue by successfully governing and developing the city.[14] As a result, the opposition-run Kampala City Council lacked the infrastructural power to maintain basic services, let alone keep up with the changing and growing demands placed on it by the population.

These issues came together in the buildup to and aftermath of the 2007 Commonwealth Heads of Government Meeting (CHOGM), an event that included a visit from Britain's Queen Elizabeth II. The pending royal visit heightened the stakes of urban governance, turning the disorderly city into a source of national shame. Recognizing the dire state of waste management in the capital, the government invested US$3 million in a program to transform urban aesthetics and clean up for the Queen.[15] This municipal facelift, focusing on main roads and other areas that would be visible to foreign dignitaries, made clear the extent of the city's garbage crisis as well as the transformative power (if not the sustainability) of concerted executive action in the city. This superficial moment of maintenance revealed the need for more profound repairs, both in the built environment and the municipality. Scandals around CHOGM contract allocations, kickbacks, and misappropriation in the wake of the meeting consolidated the popular image of the Kampala City Council as morally polluted by corruption and materially polluting due to its inefficiency and poor service delivery. Dirty and discredited, the city council was itself in need of repair.

With this political dynamic firmly entrenched and conditions in the city deteriorating, in December 2010, just months before the February 2011 national elections, which the NRM was widely predicted to dominate, Parliament passed the Kampala Capital City Authority Act. The act established a new municipal government under the auspices of the newly created cabinet position of minister for Kampala, to be appointed by the president, reversing a decade of donor-supported decentralization policies.[16] The argument behind the formation of the KCCA was that Kampala, as the capital of the nation, is too important to be left in the hands of elected officials and requires dedicated technical staff in order to properly develop. Thus, as the capital city, Kampala became subject to the national-scale sovereignty of the president and Parliament rather than the local sovereignty of the local council system that exists in the rest of the country. The new administration set out to change the face of governance in the city through structural reorganization of municipal government, a new staff, and a fresh, clean brand image and logo. The first executive director of this body would be Jennifer Musisi, a brilliant technocrat and devout born-again Christian who had made a name for herself transforming, modernizing, and cleaning up corruption in the Uganda Revenue Authority. The new slogan for the rebranded city administration, "for a better city," couldn't help but comment on the failure of previous urban regimes.

Commenting on the establishment of the KCCA, President Museveni urged, "Don't make Kampala a battle-field but keep it clean. Musisi came to rescue the situation of the City that had got out of hands of the Local Government who had failed to manage its cleanliness, potholes and public health for 25 years."[17] As Museveni's battlefield metaphor suggests, however, the authority of the KCCA was far from given. In the February 2011 elections, Kampalans elected a new lord mayor, Erias Lukwago, a lawyer and former opposition MP representing Central Kampala for the Democratic Party. Lukwago himself, in his electoral manifesto, vowed to clean up the city, tackling the garbage crisis and endemic corruption in the administration.[18] Conflict between the mayor and the executive director, the politician and the technocrat, soon surfaced and dominated local politics for the remainder of Musisi's time at the KCCA.[19] Lukwago was hugely popular among urban youth, informal vendors, and boda-boda (motorcycle taxi) drivers because he was seen as a thorn in the president's side and had a reputation for intervening on behalf of the urban poor in courts and in Parliament when reforms threatened their presence in the city.[20] In the words of Phillip Mukiibi, a key interlocutor throughout my fieldwork who was an informal plastic trader and a staunch Lukwago supporter, "He represents us poor people, especially the youth who are ever frustrated by Museveni. And he always defends us." Between the elections and taking office, however, Kampala was rocked by protests and doused in tear gas, and Lukwago's position as ceremonial head of the KCCA became increasingly untenable. He came to embody populist resistance to the KCCA in the eyes of many young people and workers in the informal sector. For the KCCA, he was the embodiment of kavuyo (chaos) and political disorder, the obstacle to development.

In the wake of the elections, the price of basic goods soared. Goods like oil, paraffin, flour, and other commodities, constituting the basis of material reproduction for the majority of the population, suddenly became unaffordable. Many attributed rising prices to economic imbalances caused by Museveni's use of state resources to buy the election and his extreme expenditures on military hardware like Russian fighter jets.[21] In March and April 2011, seizing on widespread discontent in the city, Dr. Kizza Besigye, Museveni's defeated rival in the presidential election, and his allies including then Lord Mayor-elect Erias Lukwago, called on urban residents to take to the streets, not to protest but to commute. Dubbed "walk-to-work," these demonstrations asked people to act in solidarity with what could be called "the walking class," the urban poor who could no longer afford to take minibus taxis into the city due to price hikes attributed to the rising cost of fuel. In a repressive political climate, these protests hit on a brilliant strategy, giving protestors plausible deniability (how could the police criminalize the simple act of walking to work?) and forcing the NRM to either allow momentum to build or to actively repress protests and reveal its true nature as a militarized authoritarian regime.[22]

The police response played directly into protestors' strategy. Labeling the protests riots, police responded with tear gas, rubber bullets, live bullets, and a full display of their military-grade equipment and techniques. Nine people were killed by police across the country during the protests, including a two-year-old girl killed by a stray bullet in the town of Masaka, while many more were injured and hundreds arrested.[23] Police brutality brought more people out onto the streets, and protests escalated.[24]

Drawing on the global climate of antiauthoritarian political protest sparked by the then unfolding Arab Spring, what had begun as a populist protest against spiraling food costs transformed into a postelection protest against the Museveni presidency. Over the course of two months, the focus of the protests shifted and narrowed. The central demand changed from government intervention restricting food prices to calls for regime change. Moreover, police brutality made Besigye's body the center of the protests, as he suffered beatings and arrests at the hands of the military police. This culminated on April 28, when a plain-clothes policeman smashed the window of Besigye's car and emptied a canister of pepper spray directly into his face, temporarily blinding him.[25] Martyred, Besigye left Uganda for medical treatment in Nairobi and protests petered out until his return on, not coincidentally, the day President Museveni was swearing into office. Dramatically, Besigye attracted huge crowds to greet him at Entebbe airport and to escort him the twenty-seven miles into Kampala.[26] After the intensity of the early weeks, the protests ultimately died out because of their narrowed focus on the issue of regime change and on Kizza Besigye's person, as well as the regime's overwhelming display of militarized force and successful tactic of putting Dr. Besigye and other leaders under house arrest. Walk-to-work's effect on the city was unmistakable, however. Although the walk-to-work protests ran their course in mid-2011, large-scale urban protests have become the dominant form of political participation for Kampala's urban surplus populations contesting conditions of inequality and processes of displacement.[27] At the same time, military police presence increased dramatically and permanently. Throughout my fieldwork in 2012 to 2014, police stationed tank-like armored tactical SWAT trucks and riot police at major roundabouts in the city, new security forces proliferated, and the military basis of Museveni's power and authority was on full and constant display in Kampala.[28]

Responses to the protests were also revealing of the political culture of developmental authoritarianism. Museveni labeled protestors, including Besigye and then Lord Mayor-elect Lukwago "economic saboteurs," arguing that political protest interfered with business in Kampala and risked giving the country a bad reputation that would undo decades of work of attracting foreign direct investment.[29] In the name of consolidating developmental achievements—economic growth being a critical source of regime legitimacy—the space of politics had to be restricted and protest was encoded as a crime against development. In the wake of the protests, a suite of repressive legislation regulating public gatherings and private relations passed through Parliament with the president's support. These included the internationally notorious Anti-homosexuality Bill as well as the less publicized Anti-pornography Bill, Public Order Management Bill, and NGO Bill that radically curtailed rights of assembly, narrowed the legal scope of action for all nonstate parties, and securitized the boundaries of the political.[30] This suite of illiberal legislation can be seen as a state strategy to reassert national sovereignty after decades of neoliberal governance predicated on rescaling and redistributing sovereign authority.[31] Indeed, supporters of the Anti-homosexuality Bill couched the bill as a defense of African values and national sovereignty in the face of immoral Western cultural and political neocolonialism.[32] As in the Amin era, this state-making project centers on maintaining moral hygiene, taking place on and militarizing both city streets and the bodies of women and sexual minorities.[33] Not just paperwork, these bills have been accompanied by police crackdowns on NGO offices as well as a week-long shutdown of the nation's largest independent media company, Nation Media Group, in May 2013.[34]

Having participated actively as a leader in the walk-to-work protests and advocated the strategy of making the city ungovernable in order to achieve regime change, Lukwago was bound to clash with the newly established municipal government. When he came into office, Lukwago found that his position had been rendered purely ceremonial. Municipal power—control over programs, policy, planning, and most importantly the budget—had been passed to the executive director of the KCCA. Lord Mayor Lukwago was issued with a high-end SUV and asked to play along. He refused. He instructed traders to boycott the license fees being collected by the KCCA and took his protests to court and to the streets, leading rowdy protests at Kisekka Market in October 2012 that drew large crowds of disenfranchised youth and participants in the informal economy whose livelihoods and workplaces had been rendered disposable in the new urban order.[35] After a lengthy set of hearings surrounded by controversy, Lukwago was impeached and removed from office, charged with obstructing government work and inciting violence against municipal employees. He was ultimately reinstated by the High Court but has not played a part in any KCCA activities since 2013.[36] In a 2015 interview, Jennifer Musisi stated that "Lukwago was a problem to us. . . . There has been a lot of development in KCCA during the last two years of his absence from office,"[37] identifying Lukwago himself as the problem and juxtaposing development and politics.

In the buildup to a new round of national elections in 2016, the minister for Kampala proposed a set of amendments to the KCCA act to resolve ambiguities and contradictions in the original act. The amendments included clauses that would complete the process of electoral disenfranchisement for Kampala residents, proposing that the (still purely ceremonial) mayor be elected not by popular vote but by selection among other councillors. This bill was ultimately rejected, commenters suggest, because of bad timing. Submitted to Parliament in the days before a visit from Pope Francis to the city (which triggered another round of facelifts), Museveni recommended that the bill be withdrawn to prevent embarrassing protests.[38] President Museveni and the NRM survived the walk-to-work protests, but the newly established KCCA had a daunting task on its hands to establish its authority in the context of an urban uprising and crisis of legitimacy for the ruling regime.

To clean up City Hall, the newly formed KCCA took to the streets, where it sought to establish its authority by provisioning basic services (with waste management as the first priority) and by restoring a sense of sanity and order to the city through the enforcement of trade order ordinances. Waste was not the residual effect of urban transformation but rather an active process through which transformation was achieved. Waste is not just a metaphor but also the material terrain through which urban transformation was initiated and the authority's legitimacy founded. Waste management became a vital arena in which the new authority would demonstrate its expertise and develop its capacities for urban governance.[39] The labor of maintenance would remake the city and remake the perception of the KCCA. The object of municipal maintenance is the city itself, seen as a means of ameliorating living conditions for the urban population, attracting investment, achieving security, restoring political order (by narrowly defining and self-consciously excluding politics from everyday maintenance work), and rehabilitating the image of the nation. Cleanliness and order emerged as the KCCA's guiding values, entwining governmental care for the urban population and the consolidation of the sovereign authority to expel street vendors and ambulant hawkers from the Central Business District.

Under the new regime, waste encompassed not just garbage but also people. Kampala, a KCCA official stated, was "littered with vendors."[40] Beyond metaphor, informal workers in the city were both identified as a source of dirt and disorder and themselves cast as litter. The same infrastructures of cleanliness were applied to cleanse the city of both people and trash. To enforce trade order, the KCCA established its own police force. Clad in green and yellow polo shirts, the KCCA's new enforcement officers patrolled downtown on foot and in pickup trucks, arresting vendors and impounding their goods. The majority of the officers were recruited for their previous experience as security guards working in Iraq, employed by US military subcontractors, which attests to the connections forged between battlefields and city streets by contemporary networks of global militarism.[41] While "order" appealed to the sensibilities of the city's middle classes and the interests of downtown retailers, it was perceived as cruel and unnecessarily rough by many others who asked where the KCCA expected these vendors to go and how they were expected to earn a living under the new trade order.[42] In the summer of 2012, Uganda pop star and self-declared "ghetto president" Bobi Wine released "Tugambire ku Jennifer" (Speak with Jennifer), a protest song written from the point of view of street traders asking Bobi Wine to use his celebrity to urge Jennifer Musisi to be less cruel and to consider the consequences of banning popular economic activities. The video for the song depicts actors playing KCCA enforcement officers confiscating merchants' wares, violently arresting street vendors, and disrupting everyday Kampalans' livelihoods. The song's chorus both begs the executive director for clemency and

Figure 1. A KCCA bin on Queen's Way, Katwe. Photo by Jacob Doherty, 2012.

asserts a sense of subaltern ownership and belonging in Kampala with the lines, "Please tell Jennifer that the city is ours / as the common, poor people." The song was shortly banned from the radio, but by 2013 Bobi Wine was performing the song under the KCCA logo at the Kampala City Festival.[43] As in many other African cities, municipal projects to establish order clashed with long-established livelihood strategies.[44] Jennifer Musisi quickly came to be perceived as an unaccountably harsh "iron lady" engaged in an urban land grab on behalf of the president.[45]

While order proved controversial, cleanliness had a nearly universal appeal. Even Musisi's harshest critics had to acknowledge the success of her efforts to deal with Kampala's garbage problem. The promise of cleanliness and improved waste management extended the KCCA's constituency beyond the middle class to incorporate those historically excluded from sanitary modernity, municipal service delivery, and waste infrastructure. And who wouldn't prefer passing a bed of flowers to a heap of garbage? Beautification—landscaping roundabouts and road mediums—was the KCCA's most immediately visible effort to clean the city. The KCCA was also busy behind the scenes. They set out to regulate the burgeoning private-waste management sector through licensing, environmental impact assessments, data collection at the municipal landfill, and monitoring of routine practices and equipment quality. They attempted to develop and coordinate a system of assigned zones in which different companies would be authorized to operate, although this effort proved untenable and unpopular with operators, leading to a USh1.5 billion (US$405,000) lawsuit against the KCCA in 2016.[46] The KCCA also sought to attract more foreign direct investment in the city's waste sector, organizing conferences and courting companies who could develop waste-to-energy projects at the municipal landfill as well as developing and operating a new landfill. Most significantly for the day-to-day management of the city's trash, the KCCA brought in an entirely new staff for the Directorate of Public Health Services and Environment, including five new solid-waste management supervisors charged with planning, coordinating, and overseeing garbage collection in each of the city's five divisions.[47]

Young, university-educated, and ambitious, these supervisors were dedicated to improving collection rates, extending service to underserved neighborhoods, and cracking down on illegal littering. In interviews, these supervisors explained their roles in explicitly antipolitical terms. They described themselves as working in the technical wing of the municipal government and listed the myriad difficulties that politics pose to their work. One supervisor, Jane Amenya, explained that when the KCCA tries to fine a homeowner for having open pipes connecting their toilets to storm drains, to arrest an informal garbage collector for dumping in a wetland, or to demolish unsanitary market stalls, elected officials intervene. The distinction between technics and politics is a central emic categorical difference organizing statecraft in Kampala. Exemplifying this distinction, in a 2015 interview in *The New Vision*, Jennifer Musisi stated, "I am a technical person. This was the first time I was working with politicians and it was unpleasant. I am not bothered by the politicians unless they try to interfere with my work."[48] In the KCCA's public relations, as well as in the stated objectives of many KCCA workers I interviewed, "anti-politics" and "rendering technical" are not distant scholarly analytics but explicit, nearly verbatim policy goals.[49] The distinction between the technical and the political shapes the identities and professional aspirations of municipal workers and takes physical and institutional form in the organization of municipal offices. At Rubaga Division Headquarters, for example, the technical and political wings of the government are divided into the east and west wings of the building. Populist politicians seek to protect their voters, Amenya and other waste supervisors explained, by attempting to garner votes by interfering with municipal policies and ordinances designed to bring order to the city. In the context of a semi-autocratic developmental state with a competitive electoral environment, urban politicians' role can thus be understood as the work of mediating and mitigating disposability by buffering the tensions between municipal policy and the urban population, gaining political capital by delivering services to constituents, while protecting them from displacement and what is seen as predatory regulation. For voters in the informal economy, then, resisting displacement becomes the primary promise elected officials can offer.[50]

The KCCA's technical workers, on the other hand, explicitly saw depoliticization as vital for getting their work done. This related to an ambivalent attitude toward the urban poor. Supervisors were earnestly dedicated to improving services and living conditions for residents of the city's slums and expressed their personal desires to help develop the city in broad and inclusive terms. Yet residents of poor neighborhoods were figured as homogenous communities, obstacles to rational development who needed to be educated about the dangers of waste and configured as proper users of the city's constantly changing waste infrastructures.[51] Through their planning discourse and technocratic practices, supervisors encountered and enacted the urban poor as a distant population, another technical object to be measured and supervised, at best consulted, at worst punished.[52] Solid-waste supervisors wanted to help but were quick to invoke common behavioralist tropes and generalize about the backward, wasteful, and unhygienic habits of "these poor people," as Amenya put it, who frustrated their efforts to bring cleanliness and order to "those low-income communities."

This distance was reinforced by the organization of work. Supervisors' jobs consisted of managing parish-level managers who worked with elected officials and other leaders to identify community needs as well as find ways to convince residents to participate in KCCA waste collection efforts and cease engaging in unsanitary behaviors. Supervisors oversaw the hiring of hundreds of new "casuals"—noncontract workers paid USh5,000 a day (US$2) to load garbage trucks, desilt drains, and sweep streets. The central task of supervisors' daily routine involved going "into the field" to conduct two-to-three-hour inspection tours of their division, following the routes where trash trucks had been assigned for the morning to check that the collection had in fact taken place. Constantly on the phone with trash truck drivers and parish managers, they found kinks in the municipal waste stream (trucks' mechanical problems, conflicts between neighbors, wage and other labor disputes raised by loaders) and delegated teams to straighten them out to keep the waste stream flowing (work described in more detail in part 2). They identified new unauthorized dumpsites and stationed locals as scouts in the area to dissuade their neighbors from dumping and to report those who did to the authority. They located backlogs—long-term dumpsites in residential areas and wetland fringes where decades of uncollected rubbish had accumulated—and petitioned the KCCA to deliver excavators and the other heavy-duty equipment needed to clear them out. They documented their efforts in reports, letters, and before-and-after photographs. By all accounts, these efforts proved hugely successful. The KCCA reported that in its first year it had nearly doubled garbage collection rates in the city, from 16,000 to 33,500 tons monthly, and even the authority's sternest critics acknowledged that the city looked and felt cleaner.[53] A fleet of twelve new trash trucks circulated through the city, brightly branded with the KCCA's colors and new logo, making visible this systemic restructuring of the city's waste management infrastructure and embodying the authority of garbage.

Through maintenance work in general and waste management in particular, subaltern groups are brought into the ambit of the governmental state and urban governance becomes present in the most intimate domains of urban life. Waste's sheer ubiquity and the number of different subjects participating in disposal processes ensure that regulating waste—even if this remains an always-incomplete fantasy of mastery—means regulating vast swaths of social life.[54] While establishing urban order remained a contentious issue for populations including street vendors, motorcycle taxi drivers, customers who avail themselves of the city's widespread informal sector, and residents of unplanned low-income neighborhoods (combined, these demographics make up a vast proportion of the urban population), cleaning Kampala provided the KCCA with a broad constituency and a straightforward moral position. Cleaning thus laundered the project of ordering, one that, as in many other African cities, was highly contested and unpopular.

The objective of the kinds of routine maintenance emphasized by the KCCA is to ensure infrastructural continuity: to guarantee that pipes flow, roads are passable, and waste does not pile up. In practice, ensuring continuity requires disruption. Maintenance work involves temporary closure of space while repairs are conducted. Maintenance asks the public to "pardon our dust" while work is in progress, to "bear with the short-lived inconveniences" (as posted on street signs closing road lanes to traffic), to excuse momentary dirtiness in the name of future cleanliness, temporary congestion in the name of future circulation. Establishing these exceptional zones can entail not only closure but also demolition. Clearing the way for repair, maintenance work in Kampala often involves tearing down unlicensed structures, removing antiquated infrastructures, and otherwise clearing space to establish a tabula rasa upon which an upgraded urban order and new forms of political authority can be established.

The emergence of the KCCA represents the consolidation of technocratic over political power, of experts over elected representatives. The law that brought the KCCA into being dramatically narrowed the mandate of the elected lord mayor and invested developmental authority in the presidentially appointed executive director. The indefinite suspension of democratic representation, the replacement of an elected mayor by an appointed executive director initiated by the KCCA is the institutional equivalent of the "work in progress" signposts asking residents to excuse an exceptional rupture in ordinary circulation in the name of conducting repairs, in this case, to the nation's moral standing and reputation. This exception becomes the rule, however, because repair is continuous, banal, and everyday, a permanent and constitutive feature of urban life, not an occasional aberration. While maintenance is meant to produce infrastructural stability and service continuity, the KCCA's highly self-publicized maintenance work is meant to signal a historical rupture: the overdue establishment of a government concerned with and capable of maintenance. While the prior regime, the democratically elected populist KCC run by politicians, may have collected votes, they did not collect garbage. For the KCCA, garbage served as material evidence of the moral and technical failings of previous urban governments. Democratically elected representatives had trashed the city, so by tackling waste management as its number one priority, the KCCA drew on the evidentiary authority of garbage to sanction suspending democracy. Combining the moral authority of cleaning with the exertion of sovereign power to displace existing urban forms of life, thus constituted as waste, the routines and disciplines of waste management generate new forms of politics, rule, and urban citizenship.

Discard studies researchers have explored practices and politics of repair as a counterpart to regimes of disposability and planned obsolescence. Repair both postpones disposal and materially challenges the black-boxing of disposable commodities that authorizes limited numbers of technical experts to know and act upon the products that constitute the everyday.[55] Along these lines, anti-consumerism activists argue that "if you can't fix it, you don't own it," seeking to reanimate lost practices of repair to create bonds between people and things as an antidote to disposability.[56] Similarly, anthropologists studying cultures of consumption in the global south have called attention to repair as a historically and materially variable set of practices through which commodities' "social lives" are extended while new and unexpected uses for Western goods are improvised by their consumers.[57] Ethnographic attention to repair is an antidote to reification, taking literally the imperative to attend to the socio-material processes and practices by which material things are stabilized, bounded, and maintained. Repair is held up as a critical practice that subverts the hegemony of design intentions, as well as the ongoing process of social "de-skilling" attributed to mass-production and disposability. At the infrastructural scale, others "hail the maintainers" in an effort to disrupt the hegemonic narrative of disruptive innovation that has been taken to new extremes by Silicon Valley capitalism and to point out the dramatic extent to which the maintenance and routine infrastructural labor that constitute social reproduction are undervalued.[58] But, as the KCCA's turn to maintenance illustrates, infrastructural maintenance is also the maintenance of the forms of power and political authority that infrastructure sustains. Rather than a subversive or subaltern practice, the KCCA's efforts to ground its authority in garbage illustrate that maintenance and repair can be enrolled to legitimize and materially constitute technocratic-authoritarian modes of rule. In this case, maintenance and disposability are not juxtaposed but rather mutually constitutive. As chapter 3 illustrates, maintenance can entail large-scale moments of waste making through destructive creation.

3 Destructive Creation

The New Taxi Park is one of a cluster of transportation hubs concentrated in downtown Kampala, in the crowded streets between the Central Business District—home of City Hall, Parliament, ministries, bank headquarters, and the country's highest courts—and Owino Market—the heart of the city's small-scale manufacturing, wholesaling, and informal retail and a key hub in the city's recycling and scrap economies. This part of town is always congested, gridlocked in part due to the density of taxi and bus parks (staging areas where buses big and small wait to fill with passengers to ferry across Kampala, Uganda, and East Africa) that attract a steady stream onto the area's narrow roads. Full of passengers, taxi parks were thriving business centers. Until 2012, in addition to seven hundred traders operating "lockup shops" (concrete structures used as small restaurants, clothing shops, or electronics stalls selling phones, airtime, and access to phone chargers), the New Park was crosscut by ambulant vendors selling water, soda, candy, and snacks to eat, newspapers and books to pass the time, and handkerchiefs to wipe away the sweat that comes while waiting for the final person to fill a crowded fourteen-seater bus.

In September 2012, the KCCA began work to revamp the New Taxi Park. The evening news showed dramatic scenes of police evicting traders from their shops, of municipal workers demolishing structures with sledgehammers and bulldozers, and of taxi drivers protesting that they had nowhere to go and that their customers could not find them. Just over a year later, President Museveni opened the refurbished park. Paved with smooth tarmac and marked by freshly painted parking bays, the new New Park was a radical contrast to the dust, mud, and potholes it replaced. It had no "lockups" though, and food vendors were strictly prohibited, as the KCCA sought to implement clear distinctions between retail space and transportation infrastructure.

In the meantime, while the park was under construction, a corrugated iron fence went up around the site. On one side of the fence was maintenance space: the tabula rasa upon which the city's new infrastructure was being constructed. On the other side were piles of rubble. This rubble, the concrete debris produced by the demolition of the lockups, stayed uncollected on the pavement and was soon taken over by traders. A lively secondhand clothes market popped up literally atop the material remainder of the demolished space.[1]

This episode dramatizes the dynamics of urban development in Kampala and the centrality of waste to these processes. It begins with the municipal power to evict and demolish in order to remake the city in its own image, a vision of orderly circulation, discrete zoning, and infrastructural improvement. This development lays waste to existing structures and economies, generating material and social debris. Here *waste* is clearly a verb. As in Ezekiel, a sovereign authority will "make the land a desolation and a waste" before "the ruins will be rebuilt."[2] Like their biblical counterparts, Kampala's ruins are not naturally occurring but rather are the material result of judgment, cleansing, and sovereign power. This form of wasting is often described as creative destruction, but, as Gastón Gordillo observes, this concept recognizes destruction "only to present it as ultimately creative, thereby depoliticizing it."[3] Through the redemptive logics of progress and production, the destruction of places and livelihoods is made to appear as the inevitable giving way of antiquated modes of life to make space for development. Gordillo proposes destructive creation as a conceptual alternative that identifies the value produced through ruination that "disintegrates not just matter but the conditions of sociality."[4] Extending Gordillo's work on rubble in his analysis of the production of urban infrastructures in Vietnam, Erik Harms highlights the complexity and polyvalence of the debris of new construction. Not simply the inevitably destructive consequences of spectacular new developments, heaps of rubble "also signal possibility and future-oriented action" in ways that legitimize displacement and eviction as a form of sacrifice, bind displaced communities affectively to the futures they promise, and engender new forms of political subjectivity and contestation.[5] As in Saigon, Kampala's rubble is produced not simply by greedy, corrupt, and nefarious bad actors but also through aspirational processes of development and visions of modernization that simultaneously dispossess and interpellate the displaced. In this light, the debris of the New Taxi Park shows that municipal authority involves not just the mandate to manage waste but also a license to make waste in the name of making new urban futures.

Gordillo theorizes the rubble of modernity as a sacrifice zone, noting that "spatial destruction operates in disjointed and uneven ways, destroying some places and regions more so than it does others."[6] While this formulation powerfully captures the ways that processes of wasting build on and reproduce spatial and social inequalities, it is important to recognize when, how, and for whom moments of destructive production achieve the status of sacrifice. As with creative destruction, the rhetoric of sacrifice implies a logic whereby violence is retroactively redeemed—made sacred—by the progress that it founds.[7] Before it became synonymous with industrial pollution and its health impacts for fence-line communities, the rhetoric of national sacrifice zones emerged as a way to legitimize the appropriation and destruction of Native lands in the American West by the US military for nuclear weapons testing.[8] Framing this violence and national sacrifice in the Cold War context, Native lands became disposable in the name of producing national security.[9] In Kampala, the displacement that takes place in sites like the New Park are not commemorated or memorialized in any way—the effects of displacement barely register and are not framed as publicly grievable events. These forms of waste making are accepted as routine and inevitable consequences of development, scripted as a natural step toward progress.

Because destructive creation makes multiple new spaces—both the new New Park and the market that emerged on the rubble of the old—Kampala's urban development is fragmentary rather than linear. The debris that is a byproduct of the production of maintenance and repair becomes a new space. Wastelands become open, if only temporarily and precariously, to be captured by the displaced, who build new commercial infrastructures and enact their own visions of urban development and futurity. The New Park is both typical and exceptional of this process of destructive creation in Kampala. It is typical insofar as large-scale projects to remake the city's infrastructure proceeded through eviction and displacement. It is exceptional insofar as the project in whose name these evictions were carried out actually did come to fruition in a relatively timely manner. This rarely has been the case, as two alternative examples of destructive creation illustrate.

In July 2011, hundreds of residents were evicted from their homes at Nakawa Housing Estates a few miles east of downtown. Built in the 1950s but officially condemned in the 1990s, these public housing flats remained occupied until the day they were demolished.[10] The image illustrating this story in the nation's leading independent newspaper, *The Daily Monitor*, showed a young woman with a baby clasped against her chest, leaning against the remaining wall of a half-demolished room, her belongings strewn around her as a bulldozer exits the frame, leaving her amid the rubble of her razed home.[11] The flats were torn down to make room for a new public-private satellite city development planned and funded by Comer Group, a UK-based international property development firm, with "starchitect" David Adjaye attached, and featuring "an iconic office campus employing thousands of Ugandans which will form the centerpiece of the New Kampala."[12] Promotional renderings of the project showed a ring of futuristic office towers glowing brightly in the Kampala night as well as five-star hotels, shopping malls, and high-rise apartments overlooking a manicured lush green street grid. As Filip De Boeck has argued, such images congeal neoliberal aspirations, spectrally conjuring a future in which Kampala joins the pantheon of world cities of the global south. These digital renderings, however unlikely to ever be realized, have the capacity to redefine "the content and scale of what is deemed to be proper urban existence" and, in so doing, precipitate the demolition and displacement of the structures through which ordinary Kampalans inhabit the present city.[13] Despite a presidential groundbreaking ceremony in 2013, the Nakawa redevelopment project stalled due to a series of financial crises, ambiguity over land ownership, and corruption allegations.[14]

Similarly, in July 2014, clearing the way for a planned commuter train, KCCA bulldozers razed informal markets and shack settlements in Ndeeba, a flood-prone neighborhood long established along and around the disused railway line running through a bustling area south of town.[15] Displaced residents scattered around town, finding new places to build shacks in slums, wetlands, and, in the case of one group, at a dumpsite. There, they began collecting and trading in plastic bottles as well as participating in a community outreach project run by a team of young eco-artists from nearby Kyambogo University. This land too was soon claimed by a developer looking to build on the industrially zoned swampy land. The families were subject to another round of evictions and displacement. After delays arising from legal challenges to the project's resettlement and compensation program, questions about the project's funding, and extended negotiations between the government and Rift Valley Railways (the private company that will operate the commuter rail service), the project was eventually inaugurated in the East of Kampala in 2018, but not before vendors in Ndeeba had returned to the train tracks to reestablish a market.[16] Among them was a young member of a billiards club who participated in the photo elicitation project I organized in Ndeeba; he started a business selling secondhand sneakers along the railway lines, hanging his wares from a large sunshade that he could easily insert into the marshy ground and then, equally easily, remove, fold up, and run away with on quick notice upon the arrival of KCCA trade-order enforcement officers. Making the most of the cleared space to set up a market without having to pay fees, he and other precarious vendors wary of new rounds of eviction have developed a range of similarly mobile temporary market infrastructures in order to occupy the rubble of Kampala's destructive creation.

These cases of uncreative destruction, of displacement without the ensuing development intended to retroactively legitimize eviction, reveal not only the ease with which the KCCA could exert its sovereign power to claim land but also the difficulties it faced in securing the proper connections, investments, and materials to launch new infrastructures. These delays and deferrals of development are typical of the temporality of infrastructure that Akhil Gupta identifies as suspension, the uncertain but materially and affectively productive gap between the here and now and the promised futures that infrastructures are charged with delivering.[17] They exemplify the temporal unevenness through which heterogeneous infrastructures are assembled. Infrastructures accrete gradually over time in what Sophia Stamatopoulou-Robbins identifies as the disjointed, lurching, jerky, or "choppy temporality" of infrastructure, marked by spasms of activity punctuating long periods of suspension.[18] These temporal-spatial lags, and the spaces that emerge within them, bring into view the entanglement of investment and abandonment and the actual conditions of life amid destructive creation's ongoing ruination. Here, waste is not simply a preexisting technical problem to be solved by the good governance of the new municipal authority but also the constant material effect of processes of urban renewal and repair. The foundational authority of waste management involves not just taking garbage away but also making waste.

Because the KCCA did not come into being in a vacuum of authority, establishing its authority entailed destructive creation: making waste of existing institutions, regimes of municipal power, and infrastructures. This dynamic was particularly visible in the KCCA's monthly cleaning exercises. This campaign, named Keep Kampala Clean, borrowed a form popularized by an elite civil society organization discussed in more detail in part 3. In conjunction with local leaders, the KCCA's divisional waste management supervisors identified an area to target and coordinated volunteers from partner organizations, local elected officials, teams of trash truck drivers and loaders, and enforcement officers to meet on the final Saturday of each month to sweep roads, gather garbage, clear drains, and otherwise clean up, including displacing informal roadside vendors. These ritualized moments of destruction were followed by the predictable reemergence of disorderly economic spatial practices.

A typical cleanup I attended in Nakawa Division targeted the vicinity of a bustling commercial strip where vendors were selling basic foodstuffs from tarps and kiosks in front of a row of more established retail traders in concrete shops. As we waited for a trash truck to arrive, Francis Malinga, the KCCA waste management supervisor for this area, told me that "the difficulty here is that people are rebellious: you can tell them how to handle their waste, but they cannot listen. They are too stubborn!" He

Figure 2. KCCA workers load demolished kiosks into a garbage truck at a cleanup in Mbuya. Photo by Jacob Doherty, 2013.

went on to recount having come to blows with residents who refused to bring their trash out to KCCA trucks when they passed on the main roads and continued to dump their rubbish on the same roadsides under the cover of darkness. Vendors, traders, and residents, he complained, believe that since they pay taxes, they have no obligation to participate in waste management, seeing it as the KCCA's job. The informal vendors especially irked Malinga: "They make the place so dirty, yet they reject our message of cleanliness!" Malinga identified these recalcitrant attitudes as expressions of rebelliousness and interpreted this rebelliousness as irrational stubbornness or as a residual trace of village life. This reading of noncompliance with municipal policy as a sign of backwardness is typical of the municipal gaze on the urban poor. Because they are not properly enrolled in the micro-routines of municipal infrastructure, residents of low-income neighborhoods are coded as belonging to another time.

As trash trucks arrived and loaders began to pick up the heaps of garbage that volunteers had gathered around the area, vendors hurriedly packed up their wares, stuffing potatoes and matooke into gunny sacks, folding fruits up into tarps, and heaping fish into plastic buckets. The market was quickly depopulated, although vendors had to leave behind the heavier and more unwieldy wooden tables, kiosks, display stands, and crates that made up the market's material infrastructure. Once the loaders had gathered all the area's trash, they turned their attention to these structures. Working with enforcement officers, they dismantled vendors' kiosks and threw them into the trash trucks. While NGOs regularly organize similar cleanup exercises, the KCCA's Keep Kampala Clean campaign broadens the scope of the project of cleaning, using garbage loaders and sanitation teams to enforce trade order. "Sanitation covers everything," Francis explained. "When peace fails, we have to use force!" Using garbage loaders to displace vendors and literally send their kiosks to the landfill, these KCCA cleanups directly turn informal vendors and their small-scale infrastructures to waste. In practice, cleaning and ordering become indistinct, emerging as a unified set of techniques and infrastructures that produces and manages disposability. There were no protests. Vendors sought to escape with what they could so that, as Francis acknowledged, "they can return tomorrow." The KCCA's technocratic authority could not entirely remake the city in its image: repair is an ongoing activity, something never fully or permanently accomplished and contested through everyday acts of encroachment.

4 Selfies of the State

The Kampala Capital City Authority's project to repair the image of the city involved not just a campaign of beautification but also the documentation and representation of this campaign in order to manage the appearance of the city. Managing appearances meant circulating a particular set of images of the city, of urban transformation, and of the authority itself. The KCCA used social media to circulate these images, posting photos of Kampala in transformation on Facebook, Twitter, and Instagram.[1] In addition, press coverage, cleanup events, and the visible presence of its trucks, loaders, and enforcement teams on city streets were also critical means by which the KCCA publicized its infrastructural projects and achievements. Seeking to produce a new picture of the city, these images offer a view of the KCCA's perception of Kampala. They amount to a self-portrait of the nascent municipal government. This chapter explores how the KCCA visually narrated and represented its project of urban transformation to the public. It asks what work these visual accounts of itself did in the constitution of the KCCA's contested authority and how ordinary Kampalans used images of their own to contest the KCCA's visual narrative of transformation.

The KCCA's way of picturing itself and the city does not focus primarily on ribbon-cuttings, awe-inspiring infrastructures, or catastrophic failures, the typical ways infrastructure visually enters public culture. Rather, the KCCA publicized photos of ongoing and completed work such as repairing roads, clearing drains, and managing solid waste; images promoting outreach programs in schools and the annual City Carnival; and announcements about awards bestowed on the authority, the overseas visits of the executive director, and the accomplishments of KCCA sports teams; warnings urging the public to cease littering and to stop purchasing goods from street vendors; and pleas to be patient during works in progress. Images of waste management; waste infrastructure; and the routine work of sweeping, tidying, and cleaning the city emerged as staples of the KCCA's visual repertoire. Mixed in with inspirational quotes and holiday greetings, these images portray a hardworking, technically minded, and progressively oriented authority competently carrying out its vision of urban transformation. This portrayal is based on a surprising way of making infrastructure visible, not as spectacular accomplishment but as a banal object of technical intervention.

Infrastructure scholars have focused on two moments of infrastructural visibility: inauguration and failure. While moments of failure turn analysis toward infrastructures' materiality, moments of inauguration (even if only virtual, as in the architectural renderings of Nakawa-Naguru) highlight the overwhelming and awe-inspiring aesthetic of the sublime or the spectacle. These turn analytic attention to ways infrastructure is entangled in ideologies of modernization, development, and nationalism.[2] Historians of liberal governmentality, by contrast, have argued that infrastructure's *invisibility* is a central aspect of its role in the maintenance of rule by freedom.[3] Because they remain imperceptible, the subtle ways that infrastructures guide and direct movements materially underpin the emergence of ideologies of personal autonomy and free choice that define the liberal subject in the context of modern urban mass societies.[4] Invisible modern infrastructures thus make possible, rather than constrain, the freedom of movement and the forms of subjectivity this freedom entails, while excluding racialized and gendered others.[5] But failure disrupts this invisibility, bringing the materiality of infrastructures into the center of public debate. Moments of infrastructural failure and interruption are theorized as events that disclose the materiality of infrastructures.[6] Derailments bring the properties of train tracks to the fore;

blackouts make citizens consider the minutiae of electrical grids.[7] These two moments, the sublime inauguration and the material ruin, are cast as opposites, held apart as the two temporal poles of infrastructure's life course. (Except in the minds of FEMA planners and Hollywood directors paid to imagine catastrophic collapse.) In between these eventful moments, the "infra"—the below—of infrastructure dominates, and the networked systems that sustain urban life are meant to remain out of sight and out of mind. Infrastructure, the story goes, is working when we do not notice it.[8]

This emphasis on interruption, however, is predicated on the Eurocentric assumption that infrastructures regularly function as intended, an assumption that does not hold across much of the postcolonial urban world. It also ignores the proliferation of media that does pay attention to ordinary infrastructures, making them visible and public in ways that can contest or consolidate states' authority.[9] Moments of failure and interruption reveal and reproduce inequalities of access and the radically divergent means by which distinct class groups attempt to remain connected. For the majority of the world's urban population, infrastructural interruption constitutes the normal rather than exceptional condition of life, highlighting the socially constructed and politically laden nature of the designation *crisis*. In his discussion of the colonial sublime, Larkin contrasts the invisibility of infrastructure within "advanced liberal" societies with the spectacle made of infrastructure by the colonial state. These spectacles, he argues, dramatized colonial difference, demonstrating the gap between colonizer and colonized, while simultaneously promising to deliver the modernizing development that will bridge that gap. Infrastructure was made visible within the terms of the colonial exchange whereby sovereignty was traded for technological progress. Spectacular displays marked the inauguration of dams, railroads, and electrification in order to overwhelm the senses of Britain's colonial subjects and build a sense of attachment to the futures they promised. Following independence, postcolonial governments equally relied on the spectacle of infrastructure to knit together new nation-states and set them on the path of modernization.[10] Infrastructure and its inaugural visibility have thus long been central to the ways in which colonial and postcolonial states seek legitimacy.

Infrastructure's sublime effect is ephemeral and fleeting, as are the political attachments it forms. Proximity and intimacy with technologies and infrastructures erode their capacity to inspire overwhelming awe, while time makes visible the effects of neglect and disrepair. Kampala's residents are accustomed to hearing the lavish promises of large-scale projects that are meant to bring development and secure a bright future. After decades of such promises, however, many urban residents are skeptical; spectacular images of infrastructural futures do not inspire but rather serve as reminders of past failures and the gap between Kampalans' aspirations and their present conditions. In this context, the KCCA's account of itself and its work relies on much more mundane affects than the sublime to constitute its authority. In the KCCA's visual repertoire, trash trucks replace trains, filled-in potholes replace hydroelectric dams, and parking lots replace train tracks. A May 2014 post, for example, shows a newly installed cement trash bin emblazoned with the KCCA logo, situated on a freshly swept sidewalk with a sole pedestrian (and no vendors) walking past the lush greenery of the KCCA's gardens, the solidity of the cement bin anchoring the sense of peace and stability of the urban scene. Similarly, a March 2014 post celebrating sanitation week depicts a group of women and children chatting as they wait patiently next to their sacks of rubbish, drying laundry hanging in the sun, as high-visibility yellow vest–clad municipal workers deliver a bright red dumpster to collect the neighborhood's waste. For the most part, these images evoke order, not awe; calm competence, not upheaval. One counterexample, where the KCCA did attempt to mobilize the infrastructural sublime, is instructive.

In May 2014, a story broke in the Ugandan press that the KCCA was planning to install a cable car as part of its efforts to transform transportation infrastructure in the city. A few days later, the KCCA confirmed this rumor with a post on its Facebook feed that showed a photograph of a cable car, with six gondolas hanging aloft from a towering pylon, dramatically silhouetted against a stormy sky with the caption, "Cable Cars in Kampala. A mode of transport able to operate in all sorts of weather and combat the perennial traffic jams in the City. What are your views?" An image of a cable car, surreally adorned with advertising for the game Angry Birds, also featured prominently on the cover of a strategic five-year plan that the KCCA issued in 2014, which listed cable cars as part of the plan to revamp urban transport infrastructure.[11] Rather than inspiring the awe of the "dynamic sublime," this representation of a possible future for Kampala elicited a torrent of mockery when people did share their views.[12] Posting underneath the photo, one man inquired, "What drug have you smoked this morning?" Another wrote, "What is going to power them? If it is UMEME [the notoriously unreliable national public-private power company] then we should brace ourselves to hang in space so often. I hope the cables are not stolen while we hang up there." "I will not ride unless they come with parachutes" wrote another. These posters point to the KCCA's questionable ability to assemble and sustain the resources needed to properly run and maintain such futuristic infrastructure. The very idea becomes a joke, the aspiration held up against the realities of Kampala's infrastructural present: "The biggest joke I have ever heard! In a country where people openly defecate???" Here, citizens are disaffected; they see that the emperor has no clothes, the clothes, in this case, being maintenance. Kampalans' embodied familiarity with the everyday routines of waste management and the city's uneven sanitary regime immediately discredited the speculative futures the KCCA attempted to project. Unable to provide the boring basics of urban infrastructure like sanitation, the KCCA's promise of airborne mobility evoked not legitimacy but laughter and cynicism, as everyday encounters with waste emerge as a standpoint from which to comment on and critique the state's priorities. Social media allows ordinary Kampalans to speak back to the state's poetics of infrastructure and express their disaffection, hostility, and skepticism about the official narratives and speculative futures they offer.

The cable-car imagery is in stark contrast to the bulk of the images posted to the KCCA's Facebook account, however. Rather than attempting to produce the infrastructural sublime, a task that the population—or at least the most vocal members in print, radio, and social media—is too skeptical to accept, the KCCA has used other techniques to disrupt everyday modes of perception. It does so not through publicizing photographs that inspire awe but by illustrating the production of the everyday itself by documenting banal forms of maintenance, upgrade, and repair. These images ground the new authority's legitimacy not in its production of a virtual spectacular but in its attention to the everyday and ordinarily invisible practices of government. The KCCA's visual repertoire focuses on the mundane work of urban repair, maintenance, upkeep, and upgrading. They show newly installed trash cans, street sweepers working in their newly acquired reflector jackets, and trash trucks doing their rounds to collect municipal waste. They show ongoing road repairs, drainage channels being cleared of silt, and gardeners tending to green spaces. One series of images posted in October 2014 showed a team of cleaners hunched over as they swept the street with reed brushes, wearing gumboots, gloves, dust masks, and reflective vests and pushing KCCA-branded wheelie bins to gather dust and debris. The post gave "credit to all our workers who keep our streets clean always," urged the city to "please desist from throwing trash on streets, make optimum use of the trash bins around the city," and reminded readers that "a clean City is good for all #ForABetterCity." Both the literal content and the overall tone of the KCCA's visual repertoire is "work in progress." Another October 2014 series, for instance, showed workers clearing silt and installing cement piping in the drainage channel alongside the city's ring road. In contrast to the spectral futures of architectural renderings, materiality is key to this visual rhetoric as the brute facticity of maintenance and repair take center stage. By focusing on laboring bodies, heavy rocks, heaping wheelbarrows, and construction equipment, these images provide irrefutable documentation of ongoing works and the boots-on-the-ground effort involved in upgrading the city.

Between 2011 and 2014, the dominant aesthetic of these photos was the flat naturalist realism that characterizes technocratic reportage. I observed several KCCA employees taking photos in the course of their work and asked them about it. In each instance they were slightly baffled, responding that they just wanted to capture the scene for their reports. They were slightly concerned with composition, with ensuring that a cleared road or targeted dumpsite was centrally in frame, but they did not spend much time or thought on how the photos looked, taking them quickly from eye level. Stripped of the artistic elements that could produce the sublime, the photos were meant to be purely objective representations of the KCCA's work. In early 2014, this aesthetic began to change, and a more professional set of images from a new communications team—bearing the hallmarks of a more deliberate use of photographic techniques, including composition, focus and blur, lighting, and color, as well as higher-quality cameras—started to populate the KCCA's social media feeds. Still, these enhancements did not shift the documentary style of the photography or the tone and content of the images.

As with the colonial sublime, these representations are part of an exchange. In exchange for developmental authoritarianism's antipolitical and undemocratic form of technocratic rule, the KCCA delivers maintenance, not spectacle. The production of the everyday in this configuration constitutes in itself a rupture in historical experience, disrupting residents' resignation to urban neglect and marking the distinction between the era of the competent technocrats of the KCCA and the disorderly city of the populist-elected mayors from the opposition parties. This contrast is marked most explicitly in before-and-after images that document the degraded condition of the city inherited by the KCCA and the work it is doing to upgrade urban infrastructure. An April 2014 post marked the third anniversary of the KCCA with a side-by-side graphic contrasting, on the left, an image of the flooded New Taxi Park showing vendor's stalls chaotically covered with tarps and passengers wading through knee-deep dirty water and, on the right, the newly refurbished park with orderly rows of taxis and buses, passengers walking freely on the clear paths through the park, and the Nakivubo Channel (the city's often overflowing main drain) confined to a deep trench outside the park. The KCCA, these images show, is keeping Kampala's floods at bay, ensuring orderly circulation, and upgrading the material conditions of everyday life. Marking epochal time through this progress, the caption reads, "It's been 3 years down the road and there are visible changes in the city." Similar images show potholes being sealed, dirt roads being paved, heaps of uncollected garbage accumulated over years being cleared away, streetlights installed, and pavements cleared of vendors. In these images, banal municipal labor is presented as a novelty and held up in contrast to the disruptive and riotous behavior of the city's elected mayor. Because maintenance and repair are the basis for KCCA's legitimacy—insofar as it positions itself in opposition to the failed regimes of previous urban administrations under the rule of elected members of opposition parties—infrastructure becomes visible differently. It is publicized, and public making, in order to signal a moment of historical rupture, to dramatize the difference between technocratic power and populist politics.

As chapter 3 showed, this epochal urban transformation is accomplished through processes of destructive creation that entail mass demolition of small kiosks, concrete shops, market stalls, and semipermanent structures—as well as evicting more itinerant street vendors, hawkers, and others who make a living in spaces zoned for pure circulation.[13] The KCCA does not hide this aspect of its work, posting photographs of "voluntary demolitions" on its Facebook feed, visualizing public compliance with, if not consent to, destructive creation. A slew of demolitions in September 2014, for example, was documented on Facebook in images showing heaps of concrete rubble along roadsides where shops had encroached on the road reserve and showing workers tearing down mobile money kiosks and phone-repair stands fashioned from shipping containers in scenes directly mirroring those depicted in the music video for Bobi Wine's protest song "Tugambire Ku Jennifer." Appearing alongside images of waste collection, cleanup exercises, and street sweeping, these scenes are rendered commensurable with the rest of municipal maintenance depicted in the feed. They are represented as another form of beneficent urban governance, part of the banal, purely technocratic work of maintenance, repair, and beautification. Bulldozing shops becomes just like sweeping up dust.

A Facebook feed is a somewhat open space, and citizens have taken advantage of the space opened up by the KCCA to speak up and represent the city of their own everyday experiences. These online protests take different forms. In a September 2014 post, the KCCA urged citizens to take on the responsibility of better garbage disposal, arguing that littering is the cause of clogged drains and urban flooding. "Desist from the vice (throwing garbage) in drainage channels/roads," the KCCA insisted, along with a photo of plastic bottles and other everyday detritus they had pulled from clogged drains. The images here are accusations, evidence of the failure of the population to care for the city and of the scale of the task the KCCA faces in repairing Kampala. The KCCA's visual rhetoric here illustrates the logic of citizenship under technocratic rule, defined not in terms of political rights or electoral representation but as a form of responsible conduct. It is exemplified not by engaging in political speech, for example, but by throwing trash in a designated bin, participating in the everyday project of urban maintenance. Rejecting this logic, people used the post as an occasion to redistribute blame and responsibility. Pointing to poor planning, inadequate enforcement, and changes to municipal waste collection policy that have made people less likely to receive services, they rejected the moral framing of floods as an outcome of public vice and focused attention on the structural causes of flooding.

In addition to commenting on KCCA posts, some commenters take on the KCCA at its own representational game, using the Facebook stream to post photos that contest the image of smooth circulation and beautification presented by the municipality and to demand the extension of municipal services. One photo, captioned simply "totally blocked," shows a man in a wheelchair trapped behind a row of yellow bollards and unable to cross the street. In this way, it depicts the routine thoughtlessness of planning that means the city fails to accommodate its citizens with disabilities. Similar posts register failures of service delivery or use photos to depict infrastructural neglect. Another poster shared an image of an open manhole in the middle of a pavement and asks for help, telling "the KCCA to fill this hole. . . . I get so pissed each time I pass after an old lady's phone dropped in." Borrowing the KCCA's realist objective aesthetic, this image represents the city as he sees it: still in need of repair. It contests the KCCA's narrative of epochal transformation with evidence that the terms of the exchange by which the KCCA seeks to legitimize its antipolitical governance is literally full of holes.

While these replies to the KCCA's posts critique the extent of the KCCA's infrastructural improvements and demand more services and more maintenance, others contest the KCCA's authority at a more foundational level. The jokes about the cable car can be understood as a form of defacement, naming the public secret that the state does not have the infrastructural capacity to deliver a promised good, a secret whose disavowal is enacted in the KCCA's visual rhetoric.[14] Other commenters heckle the municipality, trolling the KCCA's technical posts with demands for the restoration of the lord mayor, posting the ontological statement *"waali omuloodi"* (there is a lord mayor) in order to make present their erased political voice and to politicize the KCCA's efforts to render urban governance purely technical. These protests have not been limited to social media. In November 2013, when Lord Mayor Lukwago's appeal against impeachment was in court, his supporters took out their anger on the KCCA, attacking the unpopular enforcement officers as well as garbage collectors as they worked. In response, on November 28, 2013, the day Lukwago was ordered to be reinstated by the Court of Appeals, the KCCA itself went on strike, stating: "Since March

2013, we have been caught as pawns in the tensions of the political push and shove of the City; we have been embroiled in separate, lengthy and tiresome processes before various organs and therefore have hardly had time to do our work of delivering services to the City. The above, coupled with the ensuing political controversies and violent reactions by the public on matters relating to the office of the Lord Mayor have created a hostile working environment that has put the lives of our workers in danger."[15]

The KCCA decided to cease all service delivery until it could ensure the security of its employees. Rhetorically, the statement depicts the authority as a victim of the city's politics. Rather than describing government as a space in which politics can be carried out, or the KCCA as itself a political actor, the statement disavows the disenfranchisement that sustains the KCCA as a purely technical operator. The KCCA publicized these attacks and the effects of the interruption in service on social media. An October 2013 post shows a waste collector in the dark blue KCCA uniform with a deep wound in his head and blood running down his face being escorted away from the scene of an attack as a group of muscular trade order enforcement officers gather in the background to subdue unrest. In the KCCA's visual rhetoric, the injured body of the waste collector in this scene stands in synecdochically for the governing body, under attack in the course of everyday maintenance work by the thoughtless violence of the population but resilient and protected by the city's quasi-police force. Depicting what happened when the KCCA's trade order enforcement, along with the rest of municipal operations, itself went on strike, a December 2013 photo was captioned, "This is what the city experienced along Mini Price [a downtown street] last Friday when operations were halted." The image shows trade overflowing from the busy shopfronts onto the pavement and into the street as vendors and their wares encroach on any available space. Women's shoes neatly laid in rows on a tarp sit next to a heap of plastic sandals, and hunched vendors empty sacks full of goods and dress mannequins as pedestrians weave between the vendors, go in and out of shops, and overflow the pavements to walk down the street itself. On the right of the image, a priest clasps his bag to his chest as he strides down the street; on the left, commerce spills out of frame, suggesting its infinite expansion. This chaotic commerce, the image implies, is what the KCCA's regime of order and maintenance is holding at bay. Ironically, this image would be echoed in the following years in KCCA's feeds showing photos of designated Christmas markets in which the authority closed down streets to enable informal vendors to work, accompanied by captions celebrating holiday shopping and the population's entrepreneurial spirit. But, in 2013, images of streets overtaken by traders were produced and circulated to remind the city that the Authority was the only force separating Kampala from the total chaos and disorder that would organically emerge otherwise under the rule of populist electoral forces.

The KCCA's 2013 strike lasted only one day as the minister for Kampala ordered the municipality back to work.[16] The full-page image accompanying the front-page headline "KCCA Employees Back to Work," announcing the end of the strike, shows a KCCA worker clad in a reflector vest shoveling a backlog of litter accrued in a busy downtown street during the service suspension.[17] But, while services did resume, KCCA officials at the division level remained nervous about attracting attention. From November 2013 to the conclusion of my fieldwork in July 2014, I sought to attend a cleanup event in Rubaga Division, but none were scheduled. One of the KCCA partners in this campaign explained that Rubaga—home to the city's largest slum, Kasubi, and staunch Buganda loyalists who had been at the forefront of riots in 2007, 2009, and 2010—was too politicized and the KCCA feared that if they tried to do a cleanup, rioters would attack and burn its expensive garbage trucks, interrupting the production and publicizing of municipal maintenance.

While riots and clashes with garbage collectors certainly got the KCCA's attention, other protests took more subtle forms, disrupting the KCCA's management of appearances. In one recurrent genre of complaint, residents plant maize in the middle of dirt roads. A 2013 image submitted by a reader to the national newspaper *New Vision* illustrates one instance of this, showing a heap of dirt as tall as a passing pedestrian, with a dozen stalks of maize protruding from its top and sides, litter accruing around the bottom, and a motorcycle parked to one side.[18] The mound, composed of dirt abandoned by contractors working on the road, occupies the middle of the frame and the middle of a murram (clay soil) road in front a nearly finished three-story structure of apartments and ground-floor business. The maize here materializes the incomplete urbanization residents experience. This road, the maize plant asserts, is no longer a road. Because of the failure of municipal maintenance, the city is not a city but a crumbling façade built atop soil that residents may as well use to grow food. The maize plant inverts the KCCA's reformist agenda that blames residents' behavior for infrastructural disruption. A similar ongoing form of protest involves staging scenes of going fishing in the city's innumerable potholes. One such scene, captured in 2010, shows a man dangling a fish from a line above the muddy water pooled in a pothole at the side of a road, attracting the attention of a passing minibus taxi driver.[19] Behind him, a compatriot clasps a fish in hand to present to other drivers passing along the congested road. Here again, protesters invert the aesthetic and functional norms of urban space, pointing to the ways that badly maintained roads become ponds that invite the unruly presence of rural livelihoods and ways of sustaining oneself in the city. In doing so, these protests materialize the disjuncture that residents experience between urban life and infrastructural abandonment to embarrass the municipal government. These scenes illustrate that, as Daniel Mains's ethnography of construction in urban Ethiopia likewise shows, since "there is no city without roads," mud, potholes, and maize materially signify the state's inability to produce the progress, movement, and futures it has promised.[20] Attuned to the constitutive anxieties of urban modernizing reforms, these protests stage the spectacle of acting like backward villagers to reveal the decayed infrastructural conditions of urban life and highlight the ways that technocratic governance fails to live up to its side of the infrastructural exchange.

In addition to "seeing like a state" through the optics of planning and forms of policy-knowledge production, the KCCA is invested in repairing the image of the city, and projecting a certain image of itself, of looking like a state.[21] Writing about colonial Northern Nigeria, Steven Pierce makes the argument that although classic state projects of seeing (in his case a revenue survey) have often failed to produce the knowledge they claimed, they nonetheless have enabled the government to look like a state.[22] He contends that this disjuncture—a government that looks like a state but cannot see like one—is at the root of state weakness, the population's cynical attitude toward the state as dysfunctional, and the concomitant moral economy of corruption. The KCCA's social media visibility, this chapter has argued, is an effort to make the administration look like a state that is, specifically, functional, strong, and not corrupt—in contrast to the chaotic disorder of previous populist regimes.

"Government and all Ugandans were constantly embarrassed about the state of the city," reads a 2013 KCCA statement elaborating the authority's ambitions and achievements.[23] Looking like a state entails managing how municipal power appears to its subjects. By publicizing photos of waste management infrastructure and its mundane practices of maintenance and repair, the KCCA projects an image of itself as a purely technocratic enterprise committed to the banal and everyday work of urban governance, an image deliberately juxtaposed to the disruptive picture of politics embodied in the figure of the lord mayor.

While establishing trade order remains a contentious issue for both traders and their customers (the two making up a vast proportion of the urban population), cleaning Kampala has provided the KCCA with a broad constituency and a straightforward moral position. Garbage thus has served as a material, practical, and symbolic foundation for the KCCA's authority, signaling its mandate and its ambitions as well as the infrastructural terrain upon which its legitimacy has been established. The KCCA founded its authority by identifying and tackling a waste crisis that extended from the city's streets, drainage channels, and dumpsites to the corrupted core of the previous urban administration. As a political substance, waste is hugely generative, simultaneously the object and effect of new modes of urban governance and disposability. Waste became the ground on which a new urban order was established and a critical material symbol in both representations of and challenges to the new era of urban administration.

PART II Away

5 Para-Sites

"These ones are parasites!" Victor spat out the accusation, speckling my glasses with saliva as if only that could transmit his disgust. We were cramped together in the back row of a minibus approaching Kiteezi Landfill on the outskirts of the city. The parasites in question were the traders who buy plastics fifty kilograms (110 pounds) at a time from the landfill salvagers and sell them by the ton. Their kiosks lined the road; crammed into a middle seat, I craned my neck to get a better look through the window. As a founder of the Waste Pickers Alliance, Victor had imagined ways to empower salvagers by eliminating these middlemen. "We wanted to make a co-op, and like that, if we work equally, we would out-compete these useless middlemen." Victor spoke with relish, a smile spreading across his thin face as he recounted a plan that never came to fruition, a fantasy of a ruthless cooperative structure that could have "smashed them." "We had that idea, but no land." He laughed his signature laugh, a short, high-pitched burst that emanated from deep inside. The laugh signaled his resignation to a missed opportunity. No matter. He was on to other things, applying for a job at EnviroClean Services, a brand-new waste collection company. A few months after our trip to the landfill, I told Victor that I had interviewed Chairman Stephen Tenywa, the leader of the Kiteezi salvagers who himself harbored unrealized dreams of a co-op. "I am happy that one is there. With God's will, he will do it."

Victor Opeda was a friend and key interlocutor throughout the duration of my research. I first met Victor during preliminary research in 2010 when I interviewed him in the sweltering offices of the Waste Pickers Alliance, a quasi-union he cofounded to work on behalf of loaders employed by the city's private garbage collection companies.[1] Victor made a life for himself in and through the city's waste stream, working as an informal collector, a loader, a revenue collector, a marketer, and a consultant, as well as an activist. These experiences gave him a deep knowledge of the waste stream. Hilarious and insightful, at once deeply religious and highly skeptical, Victor was an invaluable guide through Kampala's waste worlds. The following chapters take up—and rework—Victor's invocation of parasitism in order to understand the movement of waste through the city. They describe the social worlds produced in and around the municipal waste stream in order to answer the deceptively complicated question of how things are thrown away in Kampala.

There is no such place as Away. This statement, a founding axiom of discard studies, is intended as a reminder that when we throw things away they do not simply disappear. Rather, they end up in some place: somebody's neighborhood and somebody's body. The insight that there is no such place as Away emerges from decades of activism and scholarship in the environmental justice tradition. This engaged research has documented and denaturalized the spatial distribution of landfills, toxic waste, industrial pollutants, environmental hazards, ecological risk and regulation, and injurious externalities of all sorts, illustrating the ways in which this uneven geography of disposal is predicated upon, reinforces, and materially instantiates violent hierarchies of race, class, nation, and gender. As technologies that make waste develop faster than technologies for making waste go away, more and more places are inundated by waste, Stamatopoulou-Robbins points out, and this burden is unevenly borne by different populations.[2] The assertion that there is no such place as Away is intended to make visible this geography of disposal, to refute the economic calculation of externality that casts these places aside, to insist on the existence and densely inhabited nature of these places, and to ensure that they are accounted for in the ethical, political, and environmental imagination.

In what follows, I describe the socio-material worlds constructed in and around spaces constituted as Away in order to show that, although there is no such place as Away, a lot is happening there anyway. Victor's life is emblematic of the densely lived and socially rich nature of Away. It challenges the narrow conventions of representing waste worlds as entirely abject and of waste workers as "wasted lives."[3]

Victor lived in a cramped but comfortable house with a back room for sleeping and a front room for entertaining, constructed at the back of his landlord's well-manicured garden in Gaba, a fishing village on the shore of Lake Victoria that has been absorbed into Kampala's extended urban sprawl. Sitting on a jetty watching traders unload boats full of timber harvested on the lake's islands after we had shared a giant roasted tilapia at a lakeside bar bustling with Kampalans enjoying the waterfront breeze, Victor recounted how he'd come to Kampala as a seventeen-year-old in 1995 to do menial domestic work in a "rich brother's house" and finish his schooling. In his home village in western Uganda, he told me, the school was a waste of time, so he came to the city to earn money and pay his way through secondary school, eventually finishing his A-levels in 2004. He was nostalgic for his home. We planned to visit together so I could see the hills, the cattle, and the gardens and taste the local foods and so he could spread the word of God to his people. We never made the trip, however. He was always waiting for a big payday when he would have enough to take something back to his aging parents, but problems at work blocked him.

Working as a domestic cleaner through the late 1990s, he had noticed all the plastic bottles accruing in the garbage and began to collect them, realizing he could earn some extra income by gathering bottles from around the neighborhood. After finishing school, he joined the formal waste business as a loader for Great Wastes, a now defunct company that bid for municipal contracts in the early days of municipal privatization. After a year, the management saw his loyalty and honesty and entrusted him with the job of revenue collector. In that role, he met Mohammed Nvule, with whom he would go on to found the Waste Pickers Alliance (WPA) to fight for the rights of the city's loaders. The two of them were horrified by the way that Great Wastes and rival firms treated their employees. "These companies used to just employ street kids," Mohammed explained. "They underpaid them, and they were ever drunk. It was giving waste pickers a bad name. You could just see them eating from that rubbish!" They went to the Ministry for Gender, Labor, and Social Development to explain that these were workers who were in need of protection, protective gear, better salaries, and safer conditions of work, but they found that the minister had never considered the existence of waste workers and was unwilling to take on their issues. They also tried to get companies to give their loaders IDs as proof of employment, vital documents for opening bank accounts and enabling them to try to save money. This campaign culminated in a one-day strike in 2006 when loaders parked garbage trucks on Kampala Road and left them there. "That stink made our suffering known!"

After that, however, WPA struggled to maintain itself as a membership organization and shifted its attention to the informal recycling sector, where they have tried a few different projects to form a collective of salvagers. Neither Victor nor Mohammed was especially committed to this new venture though. Victor harbored dreams of building an incinerator and running a company that would collect medical waste. In 2014, he had tried to convince his new boss at EnviroClean Services to make the investment, but he had proved reluctant. Still, Victor often told me about how this business would make his fortune. Waste fueled Victor's dreams, even as he came to a darkly cynical view of the industry.

In February 2013, Victor was working for Business Waste, a new company operating in the Central Business District (CBD). His job was to collect payments from clients at the end of every month. He described his routine: "I go to each one and get that money. I have to show them the contract, or they cannot [will not] pay. They pretend they forgot, but I ask them: Where is your rubbish? They do not have it, so they know we have done our work." Victor dressed sharply, in crisp white shirts and dark ties, his suit always hanging loosely on his slender frame. I always struggled to keep up with Victor when we walked through the bustle of the CBD as he agilely moved through the crowds, managing to neither pause nor collide with anyone. He always carried two things: a precisely kept datebook and a well-thumbed Bible, both black and leather bound.

"Once I get the money, I go to the office to give it to my boss. First, I call the loaders, so they know."

"Why do the loaders care?"

"If they go that day, the boss cannot pretend. When they know he has been paid, he has to pay them."

These are the dangers of hiring a labor organizer as your revenue collector. But Business Waste relied on Victor. He had found many of their clients for them and knew the ins and outs of the waste business in the CBD. His work was to choreograph the flows of money and waste to ensure that waste moved quickly and quietly out of the busy downtown business corridor. I asked Victor if he could introduce me to his boss. We met in a stuffy office where two bored young women sat playing solitaire on desktop computers. The boss greeted us from behind his big desk and said he'd be happy to do an interview. The company collapsed before I had the chance, however. Once the company had folded, Victor gave me the backstory. Business Waste didn't actually own (or even rent) any trash trucks. They made contracts with clients and serviced them by paying drivers on municipal trucks to make additional stops on their regular routes. They employed a small team of loaders to prepare for the drivers' arrival by gathering clients' garbage and bringing it to roadsides where they hastily loaded it. Clients never knew where the trash went.

"Isn't that corruption?"

"Yes."

Again, the laugh—this time because he knew he'd shocked me, not so much because of the lurid details of the company's operations, but with his candor.

"Oh, OK. But you didn't mind?"

"No. You know, here in Uganda . . ."

Victor didn't mind because he was paid, the driver was paid, the boss made his money, the loaders got paid, and the clients had their trash removed. The KCCA loses a bit of money on fuel, he reasoned, but they don't pay anyone enough to eat so what do they expect?

Corruption was the order of the day and the perennial talk of the town. In Kampala, the word rolls off the tongue, used to explain almost any social problem or cast doubt on the legitimacy of any institution. By naming a transgression, the concept implies an ideal pure order, since polluted by impropriety.[4] It echoes the tone of journalistic exposés of governmental malfeasance, repeats transnational discourses explaining the pathology of African states, and resounds with Victor's church's view of a fallen world needing spiritual redemption. But this term does not reveal anything new about the relationships Victor describes that ease the flow of waste through the city—loaders gleaning scrap metal and plastic to supplement their incomes, off-the-books deals between public workers and private firms, middlemen trading in scraps at the edges of the landfill. These are not corruptions but rather, in many ways, the infrastructure of infrastructure: the social worlds through which service delivery comes into being.

Repurposing Victor's denunciation of the traders dealing in plastics at the edges of the municipal landfill, I identify these worlds as para-sites: *sites* to emphasize the importance of space and location, although these are also often spatially diffuse practices rather than stable locations, and *para* to emphasize the relation of proximity, of being beside. *Beside* is a helpful alternative to the linearity and assumed hierarchies of dualistic prepositions such as *above/below* or *ahead/behind*, prepositions that, as Eve Sedgwick notes, "turn from spatial descriptions into implicit narratives of, respectively, origin and telos."[5] In this way para-sites offer a useful way of thinking about informal urban infrastructures as Derridean supplements: "necessary supports, which are the conditions of possibility of any system of knowledge, but dangerous to it because they subvert its explanatory power and sovereign claims to self-adequacy."[6] Rather than uncovering a hidden world below, positing a sturdy material base upon which an ideological superstructure is erected, or signaling residual practices bound to give way under the weight of modernization, thinking through para-sites brings into focus the multiple world-making projects that take shape alongside one another, albeit in unevenly valued and violently inegalitarian ways. Para-sites are contact zones, "social spaces where disparate cultures meet, clash and grapple with each other, often in highly asymmetrical relations of domination and subordination."[7] They exist with and alongside mainstreams, although hardly on equal terms, facilitating flows while diverting materials toward unanticipated ends. Rather than a category of place or person, however, para-sites are a mode of relation.

The term *para-site* has other resonances, of course. As in Victor's invocation, *parasite* can refer to organisms that sustain themselves at the expense of a host or to individuals who rely upon others and offer nothing in return. These biological and social definitions have great everyday currency in Uganda. Long before Idi Amin's "war of economic liberation" and the 1972 decree expelling them, this rhetoric was used to demonize Ugandan Asians, framing accusations that Asian traders were a parasitic, foreign, and disloyal community engaging in economic malpractice, blocking African economic advancement, and exploiting the African peasantry.[8] Similarly xenophobic rhetoric recurred in Kampala in "environmentalist" protests against a proposal to give parts of Mabira Forest to a Ugandan Asian family to develop into sugar plantations in 2013.[9] Not unique to Uganda, comparisons to parasitic insects were used rhetorically to construct exterminable others in the contexts of the Holocaust and the Rwandan genocide.[10]

To be clear, this is not the sense I mean when I label spaces, species, and practices like recycling kiosks, informal waste collection, dumpsites, and marabou stork ecologies as para-sites. On the contrary, my goal is to illustrate the constitutive ambiguities of para-sites. Para-sitism is not strictly pathological but both relational and relative: who is para-siting whom is never stable. Recent work in microbiology shows that the difference between parasitic and mutualistic relationships is far from clear cut and can in fact change dramatically over the lifecycles of parasites and hosts.[11] Work on the human microbiome, for example, highlights the importance for human health of a range of microbes once considered parasitic.[12] This view reframes disease not as the simple presence of injurious parasites but "as the emergent outcome of complex spatiotemporal interactions between the host immune system and the internal and external microbial environment."[13] In addition to this functional dynamic, relationships between parasites and hosts also drive evolutionary dynamics, giving rise, for instance, to extravagant displays like peacock feathers.[14] Responding to this understanding of parasitism, mutualism, and symbiosis requires moving from an essentialist to a relational and ecological view of intra-actions between economic practices, between species, and between infrastructure systems. This means no longer seeking to eradicate parasites but learning to live with them.

This relational approach to parasites offers a way to understand the relationships between so-called formal and informal urban infrastructures. In Kampala, the formal waste collection sector (both public and private) is parasitic on the informal recycling trade, for example, through loaders' ability to supplement their meager wages with a second source of income they can earn from sorting through the waste they collect in the course of doing their formal work. In their messiness, the waste stream and its para-sites neatly illustrate the incomplete, multiple, and materially heterogeneous nature of urban infrastructure and the disorderly instabilities through which Kampala's waste landscapes are sedimented. Because the waste stream often depends on rather than precedes its para-sites, the hierarchy present between a mainstream and its para-sites should be understood as an effect of the material practices of marginalization, rather than as an essential feature of the para-site. Just as research on the microbiome and its parasites requires global health practitioners to rethink the antimicrobial essentialism of public health, understanding the dynamics of para-sites should prompt city planners to reconsider the place of informality in urban infrastructure.[15]

Despite pervasive state biases that view informality as polluting, unplanned, unauthorized, and often illegal, such widespread social and economic practices are vital for constructing, provisioning, maintaining, cleaning, and inhabiting complex urban systems across the global south. Like the societies and languages of imperial contact zones, these practices "are commonly regarded as chaotic, barbarous, [and] lacking in structure," a colonial way of seeing that constitutes informality as the other to the modern infrastructural ideal.[16] The debate in development policy has been how best to harness this chaos, capturing its energy and productivity. This debate constructs the informal as a frontier of formal capital expansion, proposing various means of capture, such as land titling and microfinance.[17]

Para-sites encroach upon formal urban infrastructure, diverting municipal flows to unanticipated ends that sustain economic life for populations excluded from official visions of the city's clean future. While James Scott's influential theory of infrapolitics identifies similarly invisible everyday practices of subversion—such as humor, slow-work, and skimming—his framing positions these as infrastructural in relation to demonstrations, rallies, strikes, and other more recognizable and "properly political" actions.[18] Scott's everyday infrapolitics, in other words, are the routine arts of resistance that underpin or presage "true politics." Infrapolitics exist *below* or come *before* the politics that give them their real importance. In

Figure 3. A plastic trader's kiosk in the Kyambogo Industrial Area. Photo by Jacob Doherty, 2014.

this view, infrapolitics, these everyday modes of inhabitation and appropriation, are subordinated to the "open rebellions" they supposedly presage as proto-social movements.[19] By contrast, para-sites are neither below nor before. Insofar as they emerge not by appealing to the state for service delivery but by the autoconstruction of infrastructure and the search for retroactive legitimation of this effort, para-sites are a form of direct action through which vital services are provisioned and urban belonging, in turn, is materially and socially realized.[20] Para-sites are a form of what Asef Bayat has called "the quiet encroachment of the ordinary," or, in Solomon Benjamin's terms, "occupancy urbanism."[21] These concepts refer to the millions of small everyday acts like street vending, squatting, train hopping, electric-grid hacking, and water-pipe tapping that appropriate exclusionary urban space to create lives and livelihoods that make cities inhabitable for the urban poor.

As a form of direct action, para-sites are not necessarily conflictual. Para-sitic relations work with and alongside official systems, often with the tacit but unofficial recognition of authorities. The municipal waste stream, the official system in place for regulating the flow of garbage through and out of Kampala, simultaneously relies on para-sites to take trash away and constructs a legal and policy framework that excludes them from the city. In his ethnography of waste work in Delhi, Vinay Gidwani interprets this kind of relation as an infra-economy, something at once vital to the maintenance of urban space and the reproduction of the conditions of capitalist accumulation, while remaining on the fringes of political visibility, "seen only at moments of crisis, as an object of condemnation or reform."[22] Para-sites exemplify the social condition of disposability. Products of the everyday strategies of surplus populations to carve out livelihoods in the city, they channel toxicity, injury, and risk into some of Kampala's most marginalized people and eco-systems. Made possible by the generativity of waste and its instability as a socio-material category, para-sites are predicated on waste workers' knowledge and their creative capacities to repurpose discarded materials and identify latent exchange values. Official systems depend on para-sites but simultaneously disavow them through legal frameworks, policy agendas, and modernizing ideologies that cast them off as disposable relics.

6 Legalizing Waste

In 2000, the mayor of Kampala signed the "Local Governments (Kampala City Council) (Waste Management) Ordinance" into effect. The ordinance set out a sweeping regulatory ambition: a highly formalized vision of a hermetically sealed waste stream reaching from the municipal landfill; passing through every home, workplace, and street in the city; and extending to cover every substance in the city that could become waste. The new ordinance set out to regulate the city's waste and legally construct the geographical space of Away. It created a legal framework to regularize the privatization of municipal waste collection, outlining the distribution of responsibility for the containing, storing, transporting, and disposing of waste between the municipal government, private waste collection companies, and citizens. Waste policy is guided by the ideal of confining waste to specific locations, minimizing human contact with waste, and eliminating unauthorized engagements with the waste stream. This chapter shows that this normative vision of a hermetically sealed flow of waste away from the city is predicated on a series of exclusions that have important intended and unintended world-making effects. These exclusions construct an infrastructure of responsibility, making possible the framework for private waste management and the construction of responsible urban environmental subjects. In doing so, however, they marginalize if not criminalize para-sites, the extra-legal relations and practices that contribute critically to the flow of waste through the city.

As the normative ideal that guides municipal waste planning, the waste stream is the central organizing principle for Kampala's geography of disposal. As laid out in globally circulating best-practice policy guidelines, the ideal waste stream would take waste directly from its sources to sinks, locations identified as the proper points of final storage for the city's surplus flows.[1] Ideally, separate waste streams would channel domestic, industrial, toxic, and medical wastes away from points of generation to specifically designed facilities to treat them before diverting them to recycling plants, releasing them safely back into the environment or securing means for final disposal. The solid waste stream would also be held apart from sanitary and sewerage systems built for human waste. Responsibility for human waste is delegated to another authority, the National Water and Sewerage Corporation. In actual practice, however, this distinction is not regularly sustained as garbage constantly enters the sewer/drainage systems and human waste enters the solid waste stream, primarily due to inadequate provisioning of latrine facilities in low-income areas. Although Kampala, like many African cities, is far from meeting the ideal standards set out in official policies, there are nonetheless influential guidelines shaping municipal thinking around waste management, and they have been written into an ever-changing and often experimental legal and policy framework for the city's garbage. Urban infrastructures are constructed through the dynamic interplay of regulation and improvisation, of official policy and para-sites.[2] The 2000 waste management ordinance and municipal efforts to materialize it provide a powerful way to understand this dynamic and the normative vision of disposal that structures everyday practices and relations of power within waste worlds.

In 2013, the KCCA partnered with community-based organizations (CBOs) in each division to raise awareness about the ordinance and the provisions in place for waste management in the city. These community groups invited local residents to come and listen to representatives of the waste management team from their division, elected leaders, and NGO staff who would "sensitize" them about the issue of waste in their communities. During these sparsely attended meetings, salaried municipal and

NGO workers would enumerate the provisions of the ordinance, explain the work the KCCA was doing to fulfill its role, and lament the failures of the communities to do theirs. Addressing one such meeting in the Rubaga division, a young KCCA official told the gathered residents,

> I would like to inform this community that in KCCA we have laws about garbage. If you really follow, we are trying to enforce these laws, especially on the main roads. We are not going to be lenient at all on the matters of littering garbage or illegal dumping. We also have cases where the truck comes but the people don't respond to bring the garbage. They wait for the rain and then dump in the channels so that it flows to the road. That law is still there, and the fine is one million. But the Madam [KCCA executive director, Jennifer Musisi] does not want that. She wants total imprisonment of the culprits. I think that is when we shall stop the illegal dumping.

Invoking the 2000 waste management ordinance, he established the legal basis for a new era of punitive waste management, in which routine everyday infractions would be met not with fines but with imprisonment. He went on to recognize the limited infrastructural capacity of the division government, which had only seven trucks currently available, and he promised that more would be on the way. With them would come an end to the era of lenience. Echoing these sentiments, the area's elected councillor promised to bring an iron hand to bear on those who dispose of waste in the neighborhood's drainage channels, while lamenting the shamelessness of residents who could dump rubbish in public streets. The laws of waste were coming into force with the use of full municipal punitive power to reanimate a long-established but dormant ordinance to shore up the municipal waste stream, enforce the legal place of waste in the city, and, in turn, circumscribe the space of the para-site.

The 2000 waste management ordinance defines a typology of seven categories of waste: animal, clinical, commercial, construction/demolition, medical, objectionable, and solid. This Borgesian list, establishing categories on divergent principles, some defined in terms of substance, some in terms of the location of production, and others in terms of their effects, reveals the conceptual and practical challenges of systematically sorting out the waste stream. If, as in Mary Douglas's famous formulation, dirt is "matter out of place," then in and of itself, garbage is not dirt.[3] Rather, garbage becomes problematic as litter, pollution, nuisance, or dirt when it crosses from its designated place to other spaces of encounter.[4] The purpose of the ordinance is to establish the place of waste, to delimit it within a policy framework that contains waste matter and prevents pollution by stopping material waste from becoming symbolic dirt. As with the technical practices and discourses that, in Dominique Laporte's *History of Shit*, converted feces into sewage to purge unwanted substances, practices, and relations from the medieval city and established the authority of the state, municipal waste is a precarious material-discursive effect of power.[5] Practices that violated the spatial order laid out in the ordinance were seen as polluting and, to different degrees, criminalized. The ordinance forbids or regulates practices such as dumping, scavenging, transporting, burning, and composting—that is, practices that are identified as causing a "public nuisance," which is defined as anything that is "injurious to health or offensive to the senses, or is an obstruction to the free use of property as to interfere with the comfortable enjoyment of life or property of a considerable number of persons, or which obstructs free passage or use in the customary manner."[6] Waste matters, in this formulation, insofar as it impinges on health, aesthetics, property, and circulation. Managing the municipal solid waste stream is meant to ensure the smooth functioning of other systems of mobility and value.

The scope of the law is extreme, extending across space and into the intimate realm of the home. It goes so far as to describe the kinds of containers (plastic, sealed, portable) that should be used to store and sort waste domestically. It outlaws anything that would puncture the sealed nature of the waste stream, banning practices that may scatter, spread, distribute, or litter waste and thus threaten Kampala's health, aesthetics, property, and circulation. In some of the oddly specific practices that it bans, the ordinance is itself inscribed with the traces of subaltern ways of life. A ban on setting fires inside dumpsters, for example, is only intelligible with reference to street children's practice of lighting fires in these contained spaces to stay warm during the night. The ordinance stipulates that only municipally licensed persons and companies may interact with the waste stream, be it to collect, transport, sort, recycle, or dump waste. This process of licensing and permitting, carried out in conjunction with the National Environmental Management Authority (NEMA), was intended to legally found a new market in private waste services in order to minimize the burden of waste collection on the municipality itself.[7] As with municipalities across the neoliberal urban world, the KCC sought to transition from direct service provision to a supervisory and regulatory role within a system based on outsourcing, contracting, and cost recovery.[8]

One of the first challenges the government faced in enacting the ordinance and managing the proper place for waste in the city is knowing exactly how much of it there is to begin with. As with many other forms of statistical knowledge in African cities, data about the waste stream is unevenly produced. Many policy decisions, for example, refer to the fact that Kampala residents produce on average 0.6 kg of garbage daily, and multiply this by estimates of the population, themselves uncertain, to measure the scope of the issue. This 0.6 kg estimate appears in policy documents, municipal plans, academic studies, journalistic accounts, and interviews with officials; it forms the basis for landfill planning, collection route mapping, and infrastructural investment. I was able to trace this number back to a report produced by a German Technical Cooperation Agency study in 1990.[9] The original report was more precise; the 0.6 kg figure specified waste production in high-income areas, compared to rates of 0.2 kg and 0.15 kg per person per day in middle- and low-income areas, respectively. In 1990, when these figures appeared, the national GDP was almost 500 percent smaller than it was at the time of my research in 2013, and the explosive growth in urban construction and consumption, especially of disposable plastics, food and drink packaging, and low-cost Chinese imports, was yet to occur. While the KCCA does collect data on the amount of waste that is dumped at its landfill in Kiteezi, the denominator used to figure the percentage of waste that goes uncollected remains hard to calculate. Such radical statistical uncertainty is a constitutive feature of the municipal policy environment in the city.

Between 2000 and 2010, when it was replaced by the KCCA, the Kampala City Council experimented with different ways of managing the waste stream through devolution, cost recovery, and privatization.[10] Waste collection rates, however, remained at an estimated 40 percent.[11] Bidding for municipal contracts led to accusations of corruption and even to violent confrontations as rival firms struggled to control lucrative territory in the Central Business District. Contrary to the assumptions underpinning the outsourcing policy, private enterprises did not have more technical expertise or capacity than the municipal government, and conflicts over payments and oversight inhibited service delivery.[12] These ongoing issues came to a head in 2007 when Kampala hosted CHOGM. To prepare for this event, which included an official visit from Britain's Queen Elizabeth II, the municipality invested radically in a US$3 million urban facelift that focused particularly on the city's garbage crisis and targeting remedial waste collection efforts downtown and in areas visible from the city's main transportation corridors. The KCC spent US$200,000 on solid waste management services in the buildup to CHOGM, a supplement amounting to one-third of the annual US$600,000 solid waste management budget.[13] These improvements were short-lived, however; by 2008 Kampala was again declared to be "a mountain of garbage."[14]

By 2012, the newly formed KCCA had, for a time, reduced the municipality's emphasis on outsourcing by investing in new garbage collection equipment and letting existing contracts expire without renewal, clearing the books before seeking to initiate a new round of public-private partnerships.[15] The KCCA remains the city's largest service provider and has the highest technical capacity, as of June 2015, operating a fleet of fifty-five trucks across Kampala's five divisions. The KCCA dedicated itself to waste collection in low-income areas (described in chapter 7) and required residents of wealthy neighborhoods and proprietors of commercial premises to contract directly with licensed private collectors to take their rubbish away. A list of private service providers posted on the KCCA's website in 2013 shared seventy-six registered private companies for customers to choose from. The smallest of these companies rented their equipment on a daily basis and served highly circumscribed areas; the largest were highly capitalized, owning fleets of up to seven garbage trucks and offering coverage of the entire metropolitan area. I interviewed operations and marketing managers from fifteen of these private companies. All were candid about the fact that they did not seek to provide services in low-income areas because of the difficulty of collecting revenues. This tacit division of infrastructural labor between the public and private sectors is characteristic of what Graham and Marvin call "splintering urbanism," the fragmentation of urban services and infrastructures into ever smaller niche markets through logics of privatization that exacerbate socio-spatial inequalities.[16] This splintering is not neatly confined to the neoliberal moment, however.[17] Infrastructural dualism in Kampala has its origins in the dualist structure of the colonial state, predicated on the racialized and racializing distinction between proper urban citizens and others only ever considered temporary urban residents.[18]

The leader in the field of private waste collection at the time of research was Bin It. Founded in 1992, Bin It abstained from the competition over municipal contracts, focusing instead on building a client base in Kampala's wealthy neighborhoods, emerging elite retail spaces, and corporate business premises. Promising reliable, secure, and environmentally responsible services, Bin It modeled the collection practices and high standards to which other firms entering the market aspired.

As head of the most reputable and law-abiding company in the market, the company's director of operations, Wilson Kimera, demanded more stringent enforcement of the laws governing waste disposal. He told me that the city "should strengthen their enforcement system, because I think the industry is not as strictly regimented as you might have seen in town." Tightening enforcement on waste disposal and collection would be a key step toward a cleaner city and would reinforce the infrastructure of responsibility that his firm relies on, strengthening Bin It's competitive advantage as the most responsible firm in the industry. Victor and Mohammed from the Waste Pickers Alliance pointed to Bin It as the best model for other firms to emulate. Because Bin It provided garbage loaders with adequate protective equipment and identity cards as well as paying salaries on time, activists in the Waste Pickers Alliance used the company as a standard against which they could critique the labor practices of other companies.

The hallmarks of Bin It's services are the bright yellow stickers placed on clients' gates and the bright yellow five-gallon plastic bags that clients fill with rubbish and leave for the company to collect. These bags make it easy for clients to store waste and remove it from their compounds discretely and privately, so that loaders do not have to enter their premises. In this way, private waste collection services like Bin It's are part of the broader infrastructure that constructs and maintains the bounded space and privacy constitutive of bourgeois domesticity in the city. Through the expulsion of abject matter, this infrastructure sustains the private sphere, co-constituting home and Away.

Bin It's curbside collection is familiar to many of its foreign clients, modeled as it is on the waste practices of the suburban global north. This was an explicit point of reference for its Ugandan clients I met as well. After I explained my research project over dinner at a friend's parents' home in a posh Kampala suburb, his mother joked that if I was looking for exotic ways of handling trash, I had come to the wrong house because "we pay a company to collect it, just like you do in Europe and America." As with many elite households, however, this was only a partial picture of their waste management practices. My friend later gave me a tour of the compound to show me a pair of pigs who eat most of the kitchen scraps, a small matooke garden where other organic wastes are scattered, and a pile of ash from dry trash burned the previous evening. Identified as the culturally appropriate and "traditional" means of disposal, burning waste has been a widespread way of dealing with garbage, especially among elites with spacious walled compounds.

Bin It and other private waste companies face the challenge of converting middle-class and elite urban residents to the very idea of paying for garbage removal, a service that many have been accustomed to handling "freely" by burning, burying, or dumping it themselves or leaving this to the discretion of their domestic workers and gardeners. To do so, Bin It makes appeals both to residents' immediate affective responses to waste and to their broader conscience as responsible global citizens. A bright yellow online ad for Bin It, for instance, featured the company logo, a pot-bellied white man with an old-fashioned metal trash can slung casually over his shoulder asking, "Does garbage bother you?" It posited that "environment tops the priority of the human race in the 21st century" and stated that "safe garbage collection and disposal is our specialty in order to ensure a safe and healthy environment." Linking everyday domestic cleanliness to the fate of the human race, the ad raised the ethical stakes of domestic disposal. It positioned Bin It as capable of both producing a clean immediate environment for its clients and making a better world. Waste here traverses scales, linking everyday discomforts with pressing, almost apocalyptic, consequences.

This multi-scalar theme of responsibility runs through Bin It's advertising and the firm's presentation of itself. In another ad, for example, they referred again to the global stakes of environmentalism and explained,

"Your solution: be responsible by engaging us to ensure collection and disposal of garbage responsibly. No illegal dumping in wetlands, valleys or the vicinity. All dumping is done at the KCC gazetted landfill." Here, Bin It set itself against other, less responsible waste companies, such as Victor Opeda's corrupt Business Waste, that collect waste in town and dump it in the dumpsters distributed by the municipality, or worse, dump garbage in sensitive wetlands and streams feeding Lake Victoria, contributing to wetland degradation and global climate change. The ad called on clients to not just throw waste away but to do so responsibly. Their construction of corporate responsibility relied on a particular other—the unscrupulous waste company—and was precisely articulated in reference to the legal framework of the municipal waste stream and the sanctioned Away it constructs.

At their busy office on a narrow street where nearly every building was the office of one NGO or another, Bin It's marketing director, Ahmed Zziwa, showed me a pamphlet that their marketing agents take with them when they go door to door. It reads:

> Why allow us to collect from you?
> You prefer to relax about search[ing] for where to discard your garbage.
> You have opted to raise your family to have responsible habits to dispose waste and preserve the environment.
> You are holding a party or bash and wish to maintain your status by leaving a neat and not littered venue.
> If you BIN IT nobody bothers. If you don't everybody complains of filth. When it comes to your garbage, "BIN IT." We'll take it.

Here, Bin It draws on the moral force of personal responsibility, again articulated both in reference to immediate domestic cleanliness and a broader environmental stewardship, to promote its services. The company's services become central to the moral reproduction of proper family life and the maintenance of respectable social appearances. Maintaining cleanliness is necessary to maintaining social status, the pamphlet posits, explicitly situating responsibility within a field of interpersonal status and the surveillance of elite gossip. In this fraught image of social life, where everyone complains of filth, Bin It's customers can both relax and be responsible.

Other promotional material emphasizes that the landfill is a very safe place for clients' waste: "The security at the dumpsite is very secure and no unauthorized person is allowed entry into the dumping site with the aim of scavenging the garbage." As I describe in chapter 7, my research at the landfill shows that this is not exactly true, but the ad nonetheless constructs Bin It's responsibility in relation to the hermetically sealed waste stream. It seeks to assuage clients' possible anxieties provoked by the movement of waste from the intimate domestic space of the home to the unknown space of Away. The anonymity provided by the infrastructure of the waste stream, particularly its construction of mass waste, becomes vital for assuring a sense of security that permits clients to let their waste go.[19] In doing so, the company's promotional material disavows the existence of the informal recycling industry active at the municipal landfill. Similarly, Bin It's responsibility is never articulated in relation to its good treatment of the garbage loaders it employs. In its advertising, the clean and responsible waste stream appears as a labor-less fetish. The work of disposal once borne by the client simply disappears.

Bin It promises its clients that garbage collection enables them to simply forget about their waste, that it cannot return to haunt them as socially injurious litter, as legal problem, or as environmental threat. "We'll take it" away. Away is the basis for this form of responsibility and forgetting. Moral responsibility is made possible by a specific geography embodied in gazetted locations for collection, authorized spaces for dumping, and official municipal dumpsters and landfills, as well as by proper curbside collection methods, municipal ordinances about littering and protected wetlands, and the policed distinction between the authorized and unauthorized people who are able to access or be excluded from these spaces. Along with everyday practices and modes of governance, this state-sanctioned materiality is central to the formation of environmental subjects.[20] The KCCA's legal and policy framework thus operates as an infrastructure upon which a moral notion of personal responsibility is elaborated, materialized, produced in and through urban space, and acted on.

EnviroClean Services (ECS) was one of a few smaller companies seeking to compete with Bin It at the high end of the private waste collection market. ECS's operating and collection practices, in my observation, came closest to enacting the idealized geography of responsible disposal envisioned in the 2000 waste management ordinance. As such, they illustrated the challenges of operating strictly within the legal framework governing Kampala's waste and reveal, by their absence, the constitutive role of para-sites in the political ecology of disposal. Founded by Andrew Egunyu, a Soroti-born businessman who left Uganda to study engineering in the Netherlands, where he stayed on to work in the aviation industry for nearly twenty years, ECS entered the Kampala market in 2013. At the behest of old friends now prominent in the Kampala business community, Andrew returned to Uganda with a business plan and an idealist's urge to contribute to the development of his nation's capital city. His urge for development led him to a principled commitment to formality, to operating exclusively above board and by the books, and to refusing to pay any bribes or cut any regulatory corners. This commitment to the normative waste stream's infrastructure of responsibility, however, ultimately contributed to the collapse of the company.

Andrew ran ECS from a small office in a repurposed half shipping container situated on a dusty road across from a guesthouse in an emerging middle-income neighborhood in northern Kampala. Victor Opeda brought me to the office to meet his new boss on a rainy day in April 2013, and the three of us shouted through a conversation as rain loudly lashed the roof and walls of the office. After the downpour passed, Andrew told me that ECS had attracted capital from investors in Qatar to purchase six brand-new, modern garbage trucks. At the moment, however, they only had 150 clients, enough to send out one truck twice a week. Even then, the trucks reached the landfill only three-quarters full and the company lost money every time they sent out a truck. Adding fifty more clients in the residential areas they were already serving would make the route profitable, and Andrew preached patience.

As a marketer for EnviroClean Services, Victor was trying to expand the company's business to the suburbs on the south side of Kampala, near his home in Gaba. He needed just one big anchor client to open up the market. He targeted Speke Resorts, the plush lakeside resort owned by Uganda's wealthiest man, Sudhir Ruparelia, known simply as Sudhir. Victor knew how much plastic went to waste at Speke Resorts and wanted to set up recycling bins to "get it when it is still clean." His ambitions didn't stop there; he wanted all the resort's rubbish, knowing that they would need daily collection that would make it worthwhile for ECS to send a truck south. He would get a big commission on the Speke deal and then be able to "chase those contracts all over Makindye," the southern division full of wealthy suburbs and new developments. Sudhir was very hands on, Victor explained, negotiating such deals himself, so Victor, the former trash loader, found himself on the phone with the wealthiest man in the country. "It was tough!" he exclaimed when I asked him about it. "He is a real businessman." I was impressed. Victor took it in stride, just wanting to get the contract signed right away so he could get his commission and go visit his village. Sadly for Victor, the deal fell through when ECS was forced out of business.

Like Bin It, ECS struggled to convert residents to the idea of paying for garbage collection. The 2000 waste management ordinance was a useful resource for marketers, as it mandated that private residences subscribe to a collection service. However, ECS wanted to strike up a friendly rather than threatening relationship with its clients and so did not want to rely too strongly on municipal harassment to secure clients. ECS didn't deploy the discourses of environmental responsibility or global citizenship that Bin It used in the city's established older elite neighborhoods. Operating in the mushrooming new middle-class neighborhoods, Victor and the other marketers told me, the new trucks themselves served as mobile ads for the company. Washed every week before going out, these bright and shiny pieces of equipment attracted clients as they trundled through Kampala's under-construction residential neighborhoods. So it proved when I joined the truck team on a route one weekend. Nico, a young marketer, signed up four new clients who chased after us and cited the gleaming bright truck and the durable plastic bags distributed by ECS that they had seen neighbors using as reasons for wanting the service. "We are almost a status symbol for these people," laughed Nico, commenting on his conversation with a woman he identified as a middle-class housewife from one of the currently modest but still under construction gated homes typical of the northern suburbs ECS serves.

EnviroClean Services collection involved two teams. An advanced team of three loaders and a marketer went ahead from gate to gate, collecting the week's refuse from inside clients' compounds and piling it together at easily accessible points. The truck, with a team of two loaders and another marketer, followed and collected these piles. This method ensured that the truck stayed moving rather than delaying at each gate, a time-consuming encounter that could entail waiting for residents to unlock doors, calming aggressive guard dogs, and finding trash bags stored in obscure compound corners. Although ECS's collection technique, relying as it did on sealed bags from residences and an ever-moving truck, did not allow time for the loaders to sort through the rubbish to find plastics, loaders were able to check for scrap metal in bags that struck them as heavy but not bulky and managed to accumulate a salable sackful of mixed scrap metal in the course of a Saturday's work. Andrew was unaware of this practice, but it provided an important supplement to the loaders' wages.

Loaders developed an extensive knowledge of Kampala's domestic geographies, just as marketers relied on their ongoing studies of construction sites and the changing class landscape to identify emerging markets and potential clients. Municipal infrastructure is predicated on not only the elite forms of knowledge and calculation wielded by planners but also a diversity of modes of expertise like these. In observing these knowledge practices I found myself doing something akin to what Douglas Holmes and George Marcus refer to as para-ethnography, a form of reflexive collaboration that occurs when "the traditional subjects of study have developed something like an ethnography of both their own predicaments and those who have encroached on them, and . . . their knowledge practices in this regard are in some sense parallel to the anthropologist's and deserving of more consideration than mere representation in the archive of the world's people that anthropologists have created."[21] Although they are a long way from the technocratic and scientific forms of knowledge that Holmes and Marcus use to illustrate para-ethnography, Kampala's trash truck ethnographers are very much engaged in world-making observational practices of data collection and analysis. Monitoring new construction, calculating the economic and social profiles of new neighborhoods, mapping routes by weighing distances against road quality, their knowledge of the city is built from everyday observation and informed theorization, such as Victor's speculative mapping of routes through the city's southern extremes.

ECS's service approached the ideal enactment of the waste stream laid out in municipal policy. They moved waste from its sources in the intimate spaces of the home via sealed bags that discretely bound both waste and the domestic unit (no more than one household per bag per contract, discouraging intermingling of substances or legalities) and via compactor trucks that shed no debris en route to the municipal landfill for final disposal. Confirming this level of adherence, Andrew told me that he had turned down a tempting offer to dump waste at a nearby farm where an entrepreneurial farmer wanted to make compost. Although doing so would have saved money on fuel, it would have meant the company's records at the landfill would have come up short. Andrew feared that this disjuncture between paperwork and material practice, a deviation from the hermetically sealed waste stream, would open the company up to financial or environmental investigation.

Andrew had little patience for the KCCA. He accused them of being resentful of outsiders' good ideas and of dreaming up impracticable schemes to divide up the city territorially and assign a company to collect garbage and revenues in each district. "Kampala is too unequal for this to work," Andrew protested, dreading being assigned a low-income district that would be impossible to profitably provision. "They must do those poor areas and let us compete for those who can pay and for the [contracts to service the] markets." The KCCA was itself serving the city's major food and retail markets and collecting fees for doing so. This revenue was a major contribution to covering the city's collection costs, and the service was an important way for the municipality to exert its authority over spaces that were often politically rebellious and volatile.[22] The problem, as Andrew saw it, was that by collecting revenue for services, the KCCA was modeling itself on, and thus competing with, private businesses, working simultaneously as a regulator and competitor.[23] For Andrew, the KCCA's failure to respect the distinction between public and private was a form of unfair competition and just one of the many types of corruption he had to contend with. He complained that his tax certification, municipal registration, environmental evaluation, and import licenses had all cost his valuable time, delayed because he had refused to pay bureaucrats to have the processes "facilitated." The operating environment did not allow him to produce the clean services he envisioned.

In July 2014, Victor had found a new job working for a bank. ECS, he told me, was about to fold. Andrew's unwillingness to compromise his standards meant that the company could not profitably operate in Kampala and was facing a government investigation. Victor could not explain the specifics of this investigation, but he blamed the state's incompetence, jealousy, and paranoia. A diasporic Ugandan with ties to Gulf capital, even one working in the waste sector, Victor assured me, could be seen as a threat by the ruling regime. Though we laughed at the prospect of a rebel force taking the city riding on the backs of a fleet of garbage trucks, Victor was bitterly disappointed that the opportunities he had made for himself, especially the lucrative contract with Speke Resorts he had negotiated, would not come to fruition. Expelling the informality that makes the flow of waste possible, ECS's waste stream was too hermetically sealed from the city.

Working with the French resonances of the term *parasite*, philosopher of science Michel Serres adds the idea of static, noise, or interruption to the biological and social definition of parasitism. Drawing on cybernetic systems theory, Serres argues that noise does not simply interfere with communication but is generative, the condition of possibility for communication. Serres points out that no system "functions perfectly, that is to say, without losses, flights, wear and tear, errors, accidents, opacity"; rather, "nonfunctioning remains essential for functioning."[24] A rich ethnographic literature supports Serres's analysis. James Scott, for instance, describes work-to-rule strikes in which workers carry out their duties exactly by the book, foregoing the informal and improvisational practices that expedite production and enable circulation, and, in so doing, paralyze traffic or bring manufacturing to a halt.[25] Systems of circulation and production work, workers realize, because of, not in spite of, rule-bending. For Scott this is illustrative of a broader point that modernist state planners are ideologically biased against seeing and accounting for the informal practices that enable systems to work and, in a quest to purify systems of this polluting dirt, they produce unworkable plans doomed to failure.

For Serres, the parasite reveals the impossibility of immediacy, of nondistorted communication, and of systems that remain self-same and uncorrupted. Thus, it would not surprise Serres that waste does not flow seamlessly away from source to sink, that instead it is diverted and redirected. In Kampala, the formal waste collection sector (both public and private) is parasitic on the informal recycling trade, for example, through loaders' ability to supplement their meager wages with a second source of income they can earn from sorting through the waste they collect in the course of doing their formal work. The waste stream "works because it does not work."[26] Para-sites are, in this way, the precondition for flows.[27] Although legal orders construct a particular place for waste that differentiates garbage from pollution, the systems devised in municipal policies like the 2000 waste management ordinance work because of the practices they exclude. Such legal-spatial orders are the basis of an infrastructure of responsibility, but one that is predicated on the constitutive disavowal of para-sites, the places and practices that sustain its everyday operation.

7 Sink and Spill

Para-sites proliferate around Kiteezi, Kampala's municipal landfill. For most Kampalans, Kiteezi is Away. The final resting place for most solid waste collected in the city, the landfill has been in operation since 1996 in a peri-urban area thirteen kilometers north of City Hall, just outside the limits of Kampala City proper.[1] Ecologically, Kiteezi is a sink. Sinks are sites where wastes, toxins, and pollutants are channeled, stored, and deposited. They play vital roles in urban metabolism, processing and filtering waste before returning it, as resources of one kind or another to the environment or, alternately, keeping it sequestered away so ecologies remain in balance, not overwhelmed by waste. But not everything always remains sunk. Sinks are vulnerable to spills, moments of excess when wastes leak, leach, and breach the boundaries.[2] While waste streams are designed to make garbage go away, this flow is far from linear and subject to a variety of spills that remake, redirect, and revalorize waste. Some spills are the effects of poor planning and infrastructural neglect, as with leachate that seeps out through the ground under landfills to contaminate surrounding soils. Other spills, like the elaborate salvaging economy that captures recyclable materials and removes them from the waste stream, take work. Spills and waste streams are liquid metaphors that, like flow and leak, as

Amiel Bize points out, risk naturalizing this work, obscuring the material specificities of flows, the redistributive labor, and the ethical framing involved in making and legitimizing new resources at the margins of official infrastructures and economies.[3] This chapter considers the interplay of sink and spill at Kiteezi, examining Away as a workplace in order to illustrate the forms of life and labor constituted by para-sites and the forms of work, morality, and materiality that rechannel waste streams into recycling economies.

Trash trucks hurtle down the Lusanja-Kiteezi road on their way to the dump as drivers seek to make up for lost time on what is typically the first stretch of clear road they reach after trundling through Kampala's congestion. The trees that line the roadside are all beige and sepia, coated with a thick layer of dust from the hundreds of trucks that pass here every day. These trucks, along with the lingering odor that worsens at night and in the rainy season, are neighbors' main objection to the landfill. The trucks fill the air with dust, debris, and exhaust fumes. Occasionally they collide with pedestrians or boda-boda drivers, leading neighbors to protest by demonstrating at the landfill gates or by hurling stones at the KCCA's trucks. I reached the landfill in many different ways, riding in the cabin of private and municipal trucks, by public transport in minibus taxis, by boda-boda and on my own motorcycle. No matter the means, however, I always registered my approach by wiping grit from my eyes and feeling grime settling on my skin.

Along with schools, churches, houses, and shops, the road to Kiteezi is lined with dozens of kiosks where traders buy and sell various recyclable items including plastic, glass, metal, paper, and cardboard. Municipal and private trucks stop at these kiosks on the way to the landfill. In the course of filling these trucks, loaders sort out plastics and metals, keeping them separate in net sacks balanced atop the mounds of rubbish or tied to the roof of compactors. Drivers drop off loaders so that they can sell their materials to traders and divide the money among themselves, doubling or even tripling their USh5,000 (US$2) per day of formal wages. Drivers pause on their way back into town to pick up the loaders and continue along their routes. Traders buy four hundred to a thousand kilograms of plastics a day, depending on the capital available to them and the space they have to store it. Traders sort plastics into different types based on composition. The vast majority of the plastics involved here are empty water and soda bottles. Other common sources include broken PVC furniture, margarine tubs, TV and computer monitor casings, and car bumpers. As with the geographic knowledge produced by waste company marketers discussed in chapter 6, the intimate material knowledge of the waste stream that is produced by this way of sorting provides salvagers and traders with a deep understanding of changing patterns of consumption and production in the city.[4] I interviewed several traders standing in the shade of eight-foot-tall stacks of plastic, somehow held in place by nets and poles. These kiosks, nestled into dusty matooke groves a few yards back from the road, consist of small plots of land enclosed with tall, corrugated iron fences. Traders' most important piece of equipment is a scale, hung from tree branches, wooden frames, or hoisted above the heads of loaders. Salvagers, loaders, and traders all describe these scales as sites of contention. Each assumes that the others have rigged the scales or stuffed bottles with rocks and adjusts their calculations accordingly. "I pay 300 [US$0.12] for each kilo," one trader told me, "but then I subtract some kilos because there are always bottles with many stones [inside]." Traders know these tricks well and use them themselves when selling. During one interview, I observed a trader's employee filling bottles with water and stuffing them deep into the center of the large bundles she was preparing to weigh and sell to an investor.

While they are concentrated here, such kiosks are not confined to Kiteezi. Small-scale plastic traders have proliferated across Kampala in tandem with the growth of the market in plastic-bottled water and soda. Archetypical para-sites, these businesses exist in spaces such as railroad rights of way, obscure corners of markets, on hillsides too steep for other kinds of development, and in the half-finished rooms of home construction projects. Such plastic trading is also a viable domestic business for many Kampalan women who, with relatively little capital, can use a spare room in a compound to start a small and flexible business at home to trade on the marginal differences that accrue from dealing in higher volumes.

In March 2014, I spent a week shadowing Frank Ssonko, a municipal waste supervisor in Nakawa Division. His job was to ride in a different KCCA truck each day to observe the work and monitor the waste stream in action, reporting illicit dumpsites and consulting with the residents and market vendors that the KCCA was serving. Working from 7 a.m. to 5 p.m., we made four trips to Kiteezi each day. To reach the landfill itself, trucks turned off the main road, drove past another row of kiosks, passed through a gate and over a scale that was only sometimes in operation, and then ascended an always-under-construction murram roadway to the plateau atop the years of buried garbage where they emptied their loads. As trucks prepared to dump, groups of twenty to thirty salvagers gathered around, ready to begin the work of collecting and sorting recyclable materials. Once the trash had been emptied, these workers used rakes to spread it out to reveal plastic bottles, plastic bags, metal scrap, paper, polythene sheets, or any of the other matter they could sell at the nearby kiosks. Municipal trucks paused at this point so that the driver could get a receipt from the landfill manager confirming that they had completed one of their day's routes. Sometimes truck drivers purchased a few hundred kilograms of plastic from salvagers. Taking advantage of their now empty trucks' carrying capacity, they loaded these bottles and delivered them to processing plants in Kampala, where they could sell them for a 30–60 percent profit. Salvagers worked in an off-the-books interval after trucks dumped the garbage and before the landfill's bulldozers covered it with murram (part of the regular practices of sanitary landfill management). Their work in this temporal interval, a seam in the waste stream, was mutually beneficial. In October 2013, 438 salvagers were earning a living from the material they gathered, and the municipality extended the life of the landfill, as over fifteen metric tons of nonbiodegradable waste were reclaimed daily (according to a KCCA site engineer). This exemplifies the mutuality of the para-sitic relation: salvagers make a living on the residual value they can separate from the city's waste, and the city government is able to extract additional usage from its existing infrastructure.

Salvagers could earn USh8,000–20,000 (US$3.20–$8.00) every day, "depending on your energy," as an experienced worker put it. Because waste streams are nonlinear and entangled in complex ways with commodity supply chains and logistics, earnings also varied depending on the fluctuating and seasonal prices of different materials. The coarse and durable plastic sacks that many urban residents used to pack and dispose of rubbish, for example, arrived at Kiteezi fairly regularly with the rhythm of urban consumption of staple goods. Demand for recycled sacks, however, is seasonal as they are used to gather and distribute agricultural commodities including coffee, beans, and peas. Traders dealing in these sacks explained that they try to hold on to them until prices peak during harvests but can become physically overwhelmed by them in the meantime, with mountains of sacks stacked precariously at the edges of their kiosks.

Access to, and control over, space is vital for successful accumulation strategies and movement up the value chain in this economy, as bulky, high-volume, low-unit-price salvaged materials accrue quickly.[5] Unable to store more than a few days' worth of collection at a time, salvagers sold their collections to traders at kiosks who employed other workers to clean, sort, and pack them. Buyers representing larger investors regularly visited the row of kiosks leading up to Kiteezi to buy large quantities (never less than three tons of plastic bottles, for example) from the traders. These large investors specialized in specific materials: plastic bottles, other grades of plastic, plastic bags, paper, or metal scrap, each with its own onward destination. Plastic bottles, the most voluminous material, were taken to processing centers in the city's designated industrial areas (with high-power three-phase electricity infrastructure) where they were further cleaned, shredded using heavy machinery, packed into shipping containers, and exported to China. In 2018, as a result of China's changing policy on importing recyclable materials that sent ripples through the global recycling economy, many of the Chinese firms left the industry, and the Ugandan operations reoriented their exports toward India.[6] Many of these investors, and their representatives purchasing from traders, were themselves Chinese. Salvagers who had cared to find out (or to speculate) about the routes of the supply chain they were engaged in told me that the investors wanted the bottles to turn into fabrics, car bumpers, and new bottles—the latter something which is not actually a possibility because of the loss in quality that recycling entails, a common misconception about recycling that sustains much of the discourse around circular economies.[7]

I spent two weeks in February 2013, and again that October, conducting interviews with salvagers and traders working at Kiteezi. At the suggestion of the site manager, my research assistant George Mpanga and I met salvagers at the small restaurants where they took breaks for water, soda, and lunch. Just outside the site, through a gap in the barbed wire

Figure 4. A scrap trader's sorting yard in the Namugongo Industrial Area. Photo by Jacob Doherty, 2013.

fence, two women had set up stoves where they prepared matooke and beans in small open-sided wooden kitchens. Customers, mostly salvagers but also drivers and loaders from trash trucks, sat in the shade of trees, perched precariously on seats fashioned from discarded materials brought from the landfill: the casing of a computer monitor, broken plastic chairs, a patch of tarp stretched over a mound of cardboard.

Through the work of photographers like Sebastian Salgado, Edward Burtynsky, and Pieter Hugo, landfills of the global south have become icons of the absolute abject underside of globalization, portrayed purely as scenes of spectacular and unrelenting suffering. Perhaps naively, I was thus struck by the relaxed and jovial mood at Kiteezi. On the quiet and cool Monday morning I started work there, I found a group of young men recounting the food at a wedding some of them had attended while others debated Arsenal's typically disastrous performance in that weekend's

English Premier League matches. Salvagers I met came primarily from Kampala and from rural areas in the southeast. Finding that their "home place has no jobs," rural migrants came to Kampala to find work. Few came directly to Kiteezi, finding low-wage jobs, such as porters at construction sites, before hearing about the potential to earn more and have more autonomy as self-employed salvagers at Kiteezi. Everyone I interviewed told me that friends or relatives brought them to Kiteezi, introduced them to the site and the work, and convinced them to try it. While some salvagers had left other jobs to collect scrap to pay school fees and rent for themselves or their children, others were trying to save enough to start other businesses (salons, market stalls, animal rearing) or to buy land and build houses in their home districts. As Kathleen Millar's ethnography of Rio de Janeiro's landfill illustrates, waste work is not always a last resort defined by the absence of any alternative livelihood but sometimes a way of making a living, and forging life projects, that affords, albeit precariously, a degree of autonomy, flexibility, and sociability that allows salvagers to navigate the uncertainty and instability of violently unequal cities.[8]

"In the beginning the smell was hard for me to stand, but now you just have to get used to it," recounted a middle-aged man who had been working at Kiteezi for six years after leaving a job as a carpenter downtown. "We are used" was the common refrain when I asked about the conditions that workers experienced. In Ugandan usage, the phrase "we are used to it" is often shortened to "we are used," making the difference between accommodation and exploitation hinge on the subtlest shade of pronunciation, intonation, and context. After initial periods of adjustment to the smell, and to the hunched-over work of sorting, collecting, and carrying in the bright heat of the sun, workers settled into routines of their own design: coming to the site after completing their morning domestic responsibilities, changing into their work clothes, gathering materials until they wanted to take a break in the shade, working until they became too tired or had to return home to prepare meals or care for children, bathing onsite, changing back into their regular clothes, and walking back to nearby homes. A young woman who had just started working at Kiteezi that week told me "I am in garbage all day, and I find it uneasy to go back and touch my baby." Workers described fastidious regimes of personal cleanliness to manage such unease, bathing themselves after work at an open water tap at the edge of the landfill and again upon their return home, changing all their clothes from one space to the other, and using perfumes to mask residual odors. As with landfill workers around the world, these personal regimes reinforce potentially leaky boundaries between home and (dirty) work.[9] Salvagers keep the waste stream securely sealed Away through these techniques, even as they sustain themselves and support homes through its porosity.

I was interested in the ways that stigma, like smell, could cling to the bodies of these workers, and I tried to discuss their strategies for dealing with the shame of "working with these dirty things," as one young woman at Kiteezi put it. Some younger workers told me that they simply tell friends, neighbors, and rural relatives that they don't work, or just do odd jobs. For young unmarried people, the job can make it difficult to court, an older man explained, so they are more inclined to hide it. For people like him, with families they have supported, land they bought, and houses they had already built, the work bears little shame: "People can say what they want, but when they see what I have earned, they cannot say anything!" This was the most commonly articulated sentiment I encountered among Kiteezi salvagers. Garbage stinks, but for these workers money doesn't. Morally earned (as opposed to thievery, a distinction many salvagers drew) and morally directed (spent on families rather than on alcohol or drugs) money works as a kind of deodorant, not only uncontaminated by the dirty work of procuring it but also able to cleanse workers of the judgmental gaze of others who might despise them. This form of moral reasoning—in which the ability to feed, house, educate, and care for others is the transcendental good that validates engaging in marginal, precarious, and stigmatized labor—is common to Uganda's informal economy.[10] This marks an important contrast to the moral logic of volunteering (discussed in part 3) in which earning money from dirty work makes it a degraded and degrading form of survival, standing in contrast to the morally pure forms of citizenship enacted by unremunerated volunteers.

Dennis, a nineteen-year-old who had grown up in and around the landfill recounted, "When I joined this job? It was when my world was ended. I lost my parents, and I lacked someone to pay my school fees for me. When I came here, I got school fees, which means I've changed because

I would be out there as a thief, as a ganja guy, but now I'm working. I'm concentrated on my work where I look for my fees." Kiteezi allowed Dennis to rebuild a world for himself. Concentrating on his educational aspirations dematerialized his work to some extent. He was searching not just for plastic bottles but for school fees and a future. Other salvagers find pride in their work through the idioms of environment and development. Reflecting the growing consciousness among plastic salvagers globally about their contributions to mitigating issues like ocean plastic pollution, they point to the benefits they bring to the municipality, to the ways that their work aids in cleaning the rest of the city, and to the growth in business and construction in the areas surrounding the landfill where they are, in fact, engaged economic citizens.[11]

Aside from smell, the other major stigma that salvagers encounter is the idea that they are mad. Because many people who experience severe mental illness in Kampala and suffer multiple intersecting forms of exclusion and abandonment do, in fact, make homes in and around urban dumpsites, the two conditions are broadly conflated by many Kampalans. Spending time in, around, and intimately engaged with trash thus serves as evidence of madness. "They assume we are eating rubbish like mad people," knowingly laughed a middle-aged woman who had been working at Kiteezi for twenty years. While the landfill metaphorically feeds families via the income it generates for workers and workers are careful to dismiss rumors that they physically eat from the waste, this distinction appears to have less salience in the minds of others who stigmatize the work. "It is due to the way we dress here," explained a young woman, tugging on her T-shirt, stained a yellow-brown from hours of exposure to Kiteezi's dusty breeze. This association of madness and garbage is not limited to the landfill; salvagers I interviewed working in Kampala shared the same experience. Some of these urban salvagers play off this association for their own benefit, occasionally performing madness so that "they [onlookers] just leave us alone to do our work." Several salvagers admitted that they themselves had harbored these prejudices about the work before getting to know it. "People used to put us in the image of mad people," explained Dennis, the salvager searching for his school fees, "but we are not mad. We are people who have rules and regulations that we follow!" Here, the formalities of salvaging work, its actual orderliness, attest to its collective, regular, and rational character, defying the image of madness, a trope often used in Kampala media to describe vast swaths of urban life and the everyday practices of informal economies.

One self-imposed regulation at the landfill required salvagers to wear a uniform of blue coveralls along with protective gumboots and gloves. Salvagers strongly identified with their uniforms, seeing them as symbols of order at the landfill that dignified and legitimized their work as something organized and real, materializing their unofficial license to be there. The uniforms manifest their collective discipline and constitute the landfill as a proper workplace. They allowed them to easily distinguish between legitimate registered salvagers and potential thieves. Nonetheless, workers found their own customized ways of sporting the coveralls. Some used a belt to hold up the waist, letting the top flap behind them to stay cool, while others wore an extra protective layer over their legs to keep the coveralls cleaner. Salvagers had to start work with clean clothes and so typically had three different pairs of coveralls that they rotated, I was told—another strategy for protecting the dignity and reputation of their work. Coveralls bore evidence of another layer of formality, showing individual registration numbers on the back. These regulations, introduced by workers sometime in 2012, were aimed at controlling the number of workers on the site. Many long-standing workers complained that their numbers had increased so much while the amount of salvageable waste was actually declining due to the increasing savviness of loaders on municipal and private trucks who tried to sort out recyclables for themselves en route to the landfill. "Pioneers," salvagers who had been working at Kiteezi before the introduction of the 2012 regulation, registered for free, but newcomers had to pay a one-time fee of USh20,000 (US$8, equivalent to one to three days' work) to register. Another regulation, introduced by the Kampala City Council in 2010 at the behest of the International Labor Organization, forbade anyone under the age of eighteen from working at Kiteezi.

These regulations were enforced by Chairman Stephen Tenywa, who had been at work at the site since 1997 and had spearheaded an earlier effort to collectivize collection. A polyglot salvager, Stephen kiddingly spurned my research assistant's translation, insisting we talk in English, or French, or Swahili, or any of the other Ugandan languages he knew (and knew that I did not), asserting his own cosmopolitanism that transcended the marginal location of his work. Stephen enforced suspensions if salvagers fought, stole, or came to work drunk or dirty. He was called upon to settle disputes and to mediate between salvagers and the site managers. Relations between salvagers and the official site managers were structured by contradictions. On the one hand, salvagers provided a valuable service at no cost to the municipality, removing tons of waste that would otherwise have to be buried. On the other hand, according to the 2000 waste management ordinance, their work was illegal. The ordinance states that "the operator of a landfill shall be responsible for keeping scavengers away from the landfill."[12] Moreover, salvagers' presence violated a basic tenet of sanitary waste management, that contact between humans and the waste stream be minimized. The salvagers' work, intimately mixing their bodies and labor with unsorted municipal waste, exposed them to innumerable germs, and to dangerous debris such as broken glass and rusty nails, not to mention the longer-term dangers of breathing air laden with carcinogenic fumes.

While their presence was common knowledge for KCCA waste management staff, salvagers were not officially recognized by the municipality. As one young salvager summarized the relationship, "the KCCA minds less [does not care] about us. They look at us as those birds. They have never called us to talk about any vital thing, which I think is improper." Those birds were the nearly one thousand marabou storks—striking and ungainly birds that stand about four feet tall with seven-foot wingspans and dangling pink gular sacs—that scavenge at Kiteezi, each eating up to four kilograms of carrion, rotten fruit, and food scraps daily.[13] Except for one young man who had once been struck in the eye by a wing, salvagers did not mind working alongside the birds. "They eat their things, they don't bother us," said a salvager who had recently begun to work at Kiteezi. This was not the attitude held by the landfill's neighbors or by most other Kampalans. Giant birds have giant droppings, and these unruly circulations of waste disturb residents who park cars under trees, lay out laundry or beans and grains to dry in the sun, have matooke groves near the landfill, or unwittingly seek respite in the shade of a tree hosting a marabou stork nest. Marabou storks (their nests, bodies, and feeding patterns) are a despised companion species, another of the waste stream's contested para-sites, a hybrid techno-politico-eco niche flourishing on Kampala's rubbish, they exist in uneasy relation to urban order and cleanliness.[14]

Salvagers were registered, but this registration was initiated by their own chairman and did not entail any benefits other than limiting the number of salvagers. The number of salvagers at Kiteezi has grown with the proportion of plastics and other recyclables in the waste stream. A 1999 waste policy planning document mentions twenty-five waste pickers in the "informal private sector" who removed one ton of material from the landfill every day.[15] When I first visited Kiteezi in 2011 there were approximately two hundred salvagers working there, a workforce that more than doubled by the time I conducted interviews in October 2013. Salvagers were allowed to use a small tap at the edge of the site to clean themselves and to store their collections in nonactive areas of the landfill, but neither they nor the site managers could construct any structures that would imply permanence. To officially recognize the salvagers would be to officially recognize that the municipality was breaking its own laws and exposing workers to the dangerous matter of the waste stream, potentially exposing the municipal government, in turn, to liability and responsibility. As with the forms of leakage that Nikhil Anand describes in Mumbai's water system, Kiteezi's para-sites redirect materials beyond official infrastructures in ways that relieve certain pressures on the municipality and, as such, are tacitly accepted but remain illegal.[16] This ambiguity allows salvagers to make a living despite the law, but it equally marginalizes them, placing them in an exceptional, precarious condition beyond legality such that the entire recycling economy and the livelihoods it sustains become disposable.

In 2013, workers at Kiteezi were discussing rumors about privatization. Maybe the landfill was going to be tendered to a private company to manage. This, many feared, would mean they would be evicted from the site and replaced by a labor force they assumed would be connected by kinship or ethnicity to whoever won the contract. The one form of recognition the KCCA was offering at the time, workshops training salvagers in other skills such as making fruit juices, seemed to be preparing them for eviction. Yet as Dennis, the nineteen-year-old student, astutely observed, the Uganda National Bureau of Standards (UNBS) was cracking down on exactly this kind of small-scale, uncertified, juice production. Without the capital to secure UNBS approval, let alone secure machinery to operate at a profitable scale, he explained, juice-making was a waste of time with no future. To improve their conditions, several salvagers said they would like the KCCA, or someone else ("You, *muzungu* [white/foreign person], you can get us someone," insisted two men in their mid-twenties) to support them by buying them the machinery needed to shred plastic so they could, in the hegemonic development parlance of the day, add value to their materials themselves instead of making money for traders and foreign investors.

This uneasy relationship of nonrecognition—constituted by mutual implication in illegality, pending displacement, and vague promises of relocation—exemplifies the logic and practice of the para-site. The economy of salvage at Kiteezi, its sinks and spills, is not adequately described as either formal or informal. It is technically illegal, but it is nonetheless central to the provision of both municipal and private waste collection services. Salvagers do not have bosses and enjoy a great deal of control over the rhythms of their work, but they have little control over the material conditions of that work or the prices they are paid. Their work requires little to no capital to enter but produces value captured by large-scale foreign investors.[17] This economy generates shame and stigma but also incomes that sustain social reproduction and can override moral and affective marginalization.[18] Para-sites are not confined to the dump, however. These logics of sink and spill extend along and constitute the entire waste stream, taking place in the neighborhoods and homes where waste is gathered and the peripheral edge spaces—wetlands, empty lots, and roadsides—where waste accrues in the city.

8 Assembling the Waste Stream

Kato Mubiru loved garbage, at least that is how his constituents teased him. In addition to electing him councillor to serve and represent Kabalagala Parish, his community gave him the nickname Kasasiro (Garbage), an appellation he bore proudly. Kabalagala occupies a hillside in Makindye Division, which slopes down from the busy commercial bustle of Kibuli to a wetland that is gradually disappearing. Like many of Kampala's thirty-one slums, home to an estimated 85 percent of the urban population, Kabalagala is surrounded, and being encroached upon, by commercial and residential developments catering to the city's growing middle class and global NGO, multinational, and embassy workforces. Kato's constituency was the impoverished core of the neighborhood however, and he was launching his political career by cleaning up its rubbish.

Everyday waste management in Kabalagala became a central set of practices around which residents and their elected leaders would interact. Kato established his credentials as a hardworking leader willing to get directly involved in delivering services to his constituents. Doing the rounds of the city's dumpsites and main streets regularly made him available and integral to the community. This chapter follows Kato to describe the variety of forms of work that go into creating the waste stream in a low-income Kampala neighborhood. To deliver services, Kato's role was to know and to choreograph the area's diverse everyday domestic practices in order to assemble them into a coherent, if precarious, infrastructure. He studied domestic forms of disposal, networked with technocrats and political leaders in the KCCA, mentored young men in the informal recycling trade, monitored illegal dumpsites, recruited village health workers to bring residents out on trash days, and helped load trucks full of garbage.

Writing about waste work in Dakar, Rosalind Fredericks shows how, in structurally adjusted African cities, infrastructure is devolved onto labor.[1] In the absence of needed investment in material infrastructures like high-capacity garbage trucks, the bodies of workers and volunteers come to take the place of infrastructure, subsidizing the reproduction of the city through their own embodied labor. This chapter builds on Fredericks's insight to illustrate the heterogeneity of labor involved in the production of urban infrastructure, the importance of urban choreography in crystalizing diverse practices into a precarious system for disposal, and the unscripted role of para-sites in the constitution of municipal service delivery. All infrastructures rely on labor; they are always in the process of formation and reformation. In other words, adapting John Law's insistence on the processes and practices constitutive of structures, there is not infra-structure but infra-structuring.[2] While the 2000 waste management ordinance laid out the waste stream in law, outlining the legal and symbolic structure that defines pollution and nuisance, the waste stream is only ever realized (and even then only partially) in practice. Attending to these practices reveals that the categories of waste, and their proper places, are precarious and unstable accomplishments that must be continuously shored up. Moreover, it shows that para-sites are not antithetical to this work. As with the rule-bending norms that James Scott analyzes, it is these improvisatory para-sitic practices that sustain the normal circulation of things.[3] Nor are para-sites simply the official waste stream's shadow and constitutive other. Rather, they are central to the everyday practices of assembling waste infrastructure.

I first met Kato while participating in a voluntary cleaning exercise in Kabalagala in 2012, where he invited me to visit his parish with him. Soon after, we spent a sunny afternoon walking through Kabalagala, discussing youth unemployment and waste management, Kato's two key issues. He showed me the areas where he had had most success convincing people to dispose of their household waste responsibly. Characterizing these responsible places, Kato pointed out the well-swept front steps and walk-ways and the repurposed rice sacks stuffed full of rubbish and perched discretely against the side walls of the modest one- or two-room houses where they awaited collection. Kato explained, "We are in a slum area, but some parts are very clean. They have swept, they have put their garbage in bags. See, she [a woman in her forties who greeted Kato warmly] has collected, she has cleaned; she is doing what we want." These sacks repre-sented Kato's major accomplishment. He had convinced these residents, instead of "just throwing litter anyhow," to keep their waste neatly at home until they had a sackful to take away.

As convincing as Kato could be, his work would never be completed, due to the constant residential turnover in Kabalagala. Many residents leave as soon as they can afford to live in more secure surroundings; many others leave when personal crises force them to seek new homes in cheaper areas farther from the city center. Because the area's official waste management system relies so much on the domestic labor of its residents and their cho-reographed interaction with trucks whose arrival is unpredictable, garbage collection depends on constantly informing residents of how, where, and when to dispose of waste and persuading them to actively participate. The residential turnover, combined with changing KCCA policies, meant Kato felt he was always starting anew in assembling the waste stream.

Kabalagala attracts a steady stream of new arrivals from the "deep vil-lages" who, in the eyes of planners, "import their culture to urban cen-ters of throwing away waste in the backyard."[4] While Kato was concerned with practically coordinating collection, other campaigns have had much more ambiguous goals of raising awareness, the assumption being that "an informed public can do much to improve the effectiveness of munici-pal waste programs."[5] *Sensitization* has been the term commonly used by government agencies, politicians, and NGOs in Uganda to refer to aware-ness raising and behavior change interventions. Poor Kampalans, in this view, have been simply uninformed about proper waste management, and good new information can replace bad old village habits. This paradigm of behavior change via awareness raising that locates infrastructural gaps in the mentalities and attitudes of the urban poor is a pervasive element in waste policy research and reform efforts. Research showing that the majority of Kampalans are unwilling to pay for waste collection, for example, is used to conclude that "attempts must be made to improve willingness to pay for waste management services in the city. To achieve this, the government should concentrate first on awareness campaigns about the consequences of waste mishandling and benefits of payment for proper waste management."[6] Such sensitization campaigns have sought to shore up the waste stream by insisting, "discipline yourself" and "drop the culture of littering."[7] In practice, this commonly has amounted to campaigns that shift the responsibility for social and environmental problems onto the poor and responsibilize them for service delivery.

This assumption has been highly misguided. Residents of Kabalagala and other low-income settlements could recite, and often did in casual conversations I had about my research, the core principles of municipal solid waste management, the legal distribution of roles for managing waste, and the moral importance of households and individuals managing their waste responsibly. The issue has clearly not been a lack of awareness. As the range of strategies used to actually deal with rubbish suggested, residents were keenly aware of the city's waste infrastructure, the problems that garbage posed, and the ways they could take advantage of the city's hydrology and topography to dispose of waste in ways that serve their households, if not the city as a whole. Residents were aware not only of garbage as a socio-environmental problem but also of the normative script of municipal solid waste management, even if not everyone put these norms into practice.

One CBO I worked with during my research, RUSENCO (Rubaga Sensitization Community Organization), was formed to conduct waste management sensitization campaigns in low-income areas in Rubaga Division on behalf of the KCCA. Their main activity was organizing community meetings where they sensitized residents about waste management by reading the 2000 ordinance, translating it into Luganda, explaining its significance, and bringing local elected leaders and technical staff to answer questions. Several such meetings I attended were, however, hijacked by residents who demanded to know why the division's mayor or other representatives present at the meeting failed to hold landlords responsible for constructing adequate latrine facilities. As lacking as the area was in solid waste management services, residents' priority, in this instance, was on human waste. Far from passive uninformed vessels for awareness, they used the occasion of being sensitized to flag their own issues and insist on another distribution of responsibility.

Kato went on to show me Kabalagala's Away, the fringes of the wetland where residents had been dumping rubbish for as long as he, born and raised here, could remember. "We want to stop them from dumping here, but until we give them another way, this is what we encourage." For the time being, he tacitly endorsed the illegal para-site that contributes to helping keep Kabalagala's waste manageable. Residents either brought their trash here themselves (a chore generally given to young boys) or paid an informal waste collector to carry it away for them. These informal collectors, operating with wheelbarrows or modified bicycles, circulated through the neighborhood charging USh100–200 (US$0.04–$0.08) to pick up sacks and take them away. Kato told me that he knew most of these guys (they were all men) and wanted to find ways of supporting their business. The KCCA, on the other hand, was opposed to this form of collection and was trying to arrest and fine informal collectors, impounding their meager tools. Despite the fact that these were the only entrepreneurs who had successfully constructed a functioning "cost-recovery" waste collection service in the city's poorest areas, they were seen as polluting the city's aesthetics, as contributing to wetland degradation (if they dumped their collections in wetlands), or as taking advantage of municipal services (if they dumped at known KCCA collection points).

By 2018, the KCCA had further criminalized this kind of work as it privatized waste collection in low-income settlements, granting monopolies over set territories to private waste firms who, claiming to be acting on behalf of the KCCA, hounded and arrested informal waste collectors and impounded the simple equipment they used in their work. Rather than criminalizing this work, Kato envisioned linking it up with the official waste stream, constructing easily accessible "garbage banks" (as opposed to the hard-to-reach wetland dumpsite) where informal collectors could dump trash for the KCCA to collect on a regular basis, a vision akin to projects advocated by waste pickers in, among other places, Ethiopia,

India, and Brazil and adopted in several cities.[8] Unfortunately, because it entailed more points of contact between the population and waste, this plan to recognize and integrate para-sites into official infrastructures ran counter to the image of the hermetically sealed official waste stream, the ideal through which planners came to see informal waste collectors as "dirtying the place," in Kato's words.

The land at the dumpsite, Kato explained, was full of holes dug "some years back" by brickmakers who converted the thick clay soil into building materials for the neighborhood. These holes have been filled in with rubbish, and in turn this has helped fill in the wetland, firming up the ground sufficiently for a few bold people to construct one-room homes on this flood-prone land. "In five years you will see those big investors constructing on this land," guessed Kato. "This is what they do in the Netherlands, isn't it?" he went on, situating Kabalagala in a long and geographically diverse history of land reclamation. The extent of this hybrid process of terra-formation was evident in the steep hillside adjacent to the dumpsite. Eroded by heavy rains, we could see a cross-section of the sedimented layers of soil and plastic that constituted the hillside, plastic bags seeming to spew forth from the earth below the foundations of houses now precariously balanced on the slope.

The local dumpsite is working for now, Kato told me. The trouble is that all the trash is mixed together. "People don't know how to sort garbage. They put feces, they put condoms, they put everything there. They don't know that this one is useful, and this one, but they have mixed this. Even some of these bags, they don't sort out, they have plastic materials. It's just mixed." Echoing NEMA's vision of segregated waste streams, Kato wanted to teach people to sort their waste at home, but this sorting was not infrastructurally supported and so did not catch on. Like many other officials I met, Kato saw sorting waste as vital to improving waste collection. From this position, sorting is a mindset that needs to be inculcated in residents through sensitization, but insufficient attention has been given to the material, infrastructural, and economic coordination required to realize waste's residual value. Sorting took place nonetheless. As was the case at many of Kampala's wetlands, a trader had opened a small kiosk at the edge of the swamp, where he bought and sold plastic bottles from salvagers who mined the dumpsite for materials they could sell. "They are suffering too much," in Kato's words, because disgusting things are mixed in with the useful things and the work is needlessly dirty and degrading. Kato wanted to support the salvagers however he could. His idea was to start some youth projects in these marginal wetland para-sites. He envisioned an urban farm with a piggery where waste could be put to use to generate income for the unemployed young people of the parish. Hopefully, he exclaimed, "No waste to waste! That is my slogan."

A few months after we first met, I ran into Kato at City Hall and he invited me back to Kabalagala to see the progress he was making. We met the next Sunday so I could join in the weekly collection he had worked to secure from the KCCA. "I wrote three letters to them and finally went to those offices, and they agreed to send me a truck," Kato told me. "Now they come and collect free of charge every Sunday, when people are home." Although the KCCA did not charge residents for garbage collection, Kato pointed out that sometimes loaders solicit small charges of USh500 (US$0.20) that dissuade residents from using the service. "My people are broke! If you skip lunch, how can you pay for rubbish?" he asked, before elaborating the repertoire of practices his constituents could turn to instead of waiting patiently for the weekly visit from the trash truck. The dumpsite was still an option, though the KCCA had placed scouts there to arrest anyone caught dumping. Easier, during the rainy season, was to place sacks of rubbish near drains knowing that heavy downpours wash trash away. Other residents hid under the cover of darkness, throwing their trash onto roadsides in the middle of the night. Even as they recognized these techniques as socially, environmentally, or hygienically problematic, Kabalagala residents regularly identified these techniques to me as their primary means of disposing of waste, shrugging that they worked during the day and missed the short window for municipal collection so had no better option. Burning rubbish was harder in the crowded neighborhood, as it could lead to conflict with neighbors, but many residents burned their rubbish nonetheless. Participants in a photo-elicitation interview project I conducted with residents of other low-income areas took photos of burning rubbish and ash heaps and told me that these were their ideal best-practice for waste disposal, explaining that burning reduced the volume of waste and prevented waste from getting rotten and smelly. Although they recognized that the smoke could be unpleasant, and even toxic, they were more concerned (as are municipal waste policy makers and environmental NGOs) with the short-term health threats (malaria, diarrhea) and visual-olfactory aesthetics of rubbish than with the risks of exposure to cancer-causing dioxins and asthma-inducing particulate matter that come from burning.

Away, in Kabalagala, could be anywhere outside of the immediate area of the home: the drain, the swamp, the street, or, in the case of burnt rubbish, the air. Responding to his constituents' dissatisfaction with living in too close proximity to their waste, Kato Mubiru was trying to put more distance between his voters and Away to undo the routine of embodied disposability that residents in infrastructurally abandoned neighborhoods contend with. As an alternative to the extant practices, Kato worked hard to bring services to his constituents reliably, building the trust that a truck would come every Sunday so they could have the patience to store their trash at home. He worked to secure municipal services and to link up the neighborhood's para-sites to the official waste stream so they could properly throw their garbage away. This work produced, although incompletely, a new geography of waste, relocating Away from Kabalagala to the municipal landfill at Kiteezi and the marshy edges of Lake Victoria.

Plastered on a wall facing the busy road past Kabalagala was the common command "*TOYIWAWANO KASASIRO* FINE 100,000" (Don't dump rubbish here, fine 100,000 [US$40]). Kato told me that he had written this here months earlier, trying to curb residents' habit of bringing their domestic waste here to the side of a busy road where, about five years earlier, the KCC had placed a skip. The KCCA removed all the skips in the city (apart from those at markets) as part of its waste-management reforms. The head of the KCCA's Directorate of Public Health Services and Environment explained that the skip system caused too much littering, as residents dumped their garbage around instead of in skips and that skips routinely overflowed because the municipality did not have sufficient resources to empty them often enough. The shift from the skip to the self-loading system (described below) thus entailed a new physical distribution of garbage, with homes replacing public spaces as the temporary holding areas for domestic waste as garbage was pushed further out of sight.

The phrase *toyiwawano kasasiro*, paired with a threatening fine, often appears on walls near where skips used to be, inscribing the history of

Kampala's waste management infrastructure on the city's walls. Residents, landlords, tenants, municipal workers, police, and politicians put up this injunction, attempting to confine the waste stream to its proper place and time and to preclude the emergence of para-sites. I struggled to find out who had the authority to actually collect the fines; Kato told me that he would never do any more than threaten the fine because as an elected official he was reluctant to punish his voters. Others told me that a property owner could, if he caught someone dumping, take the offender to the local chairman (a position mandated with quasi-juridical conflict resolution duties) or to the police. The KCCA was employing scouts to observe and police dumping practices in low-income neighborhoods, and these scouts could arrest residents for dumping and threaten to take them to the KCCA's court at City Hall. On the day we met, a heap of garbage bags was against the wall. Kato and I joked about this, but he explained that this was okay on Sundays because it made for an easy point for the truck to stop and collect rubbish. The para-site had been granted a temporary stay of execution in order to facilitate the outward flow of waste.

Clad in a black NGO-branded T-shirt, Kato had spent the early morning since sunrise walking through Kabalagala with members of the area's Village Health Team (VHT), youth from a CBO that he supports, and a megaphone mobilizing his people. Walking through the areas where the KCCA trash truck was scheduled to stop, he alerted residents to the impending visit, telling them to sweep their homes, gather their trash, and be ready to bring it out to the roadside for collection. Having mobilized the neighborhood, Kato and I headed to the nearby Total Petrol Station where the KCCA's trucks were fueling up for the day. Over the previous year Kato had built up good relationships with the Makindye Division waste supervisors. "You buy them some airtime, then they can call you when a truck is ready," he told me. Even so, he went on, "I come here to guarantee that the trucks come, or else they can say 'the trucks are delayed' and then make three trips instead of four and pocket that money for fuel." The truck he had been assigned for the day was an old flatbed lorry painted green and yellow. Kato and I joined the driver in the cabin, and four loaders in blue KCCA coveralls clung onto the back of the truck as we headed into Kabalagala. Like all loaders working for the KCCA these loaders were all men. The KCCA hires older women, often widows identified by local LC1 chairmen as particularly vulnerable or needy, as street sweepers. Both loaders and sweepers are paid USh5,000 (US$2) daily and are classed as "casuals"—workers with no contracts or security of employment. This gendered division of labor relies on ideologies that naturalize sweeping as women's work and, as Samson and Miraftab have shown in South Africa, in the context of privatization, exacerbates uneven wages, union representation, and labor rights between men and women working in waste management.[9]

The truck turned down a steep and narrow dirt road, lined on either side by deep storm drains. A loader jumped down to help guide the driver and ensure that we did not crash into any of the little shops built right up to the roadway. We stopped every hundred yards or so, and Kato, who had been going ahead, would return to the truck with twenty to thirty residents, almost exclusively women and children, their rubbish dragging behind them, clutched against their chests, or balanced on their heads. Residents, with help from Kato and the CBO youth, passed their sacks up to loaders, standing on the growing heap of trash in the flatbed truck, who emptied out the sacks and returned them to their owners. Other loaders sorted through the rubbish for plastic bottles or any other recyclable materials they could gather and sell later in the day.

By 2014, this mode of collection, given the name "self-loading," in which a municipal truck drives through a neighborhood and residents are expected to bring their rubbish to the truck when it passes, had become the standard practice of the KCCA in low-income residential areas. Infrastructure here is devolved onto not only the low-paid labor of workers but also the unpaid domestic labor of the city's women and children.[10] Self-loading relies on and privileges normative, respectable gender roles in which a woman is "left at home" to tend to the house and family during the day, when the trash truck might arrive. During the many collection routes I observed on municipal garbage trucks, I never observed an adult man other than Kato and his team "self-loading" garbage, a chore relegated to women, children, and domestic workers. This mode of collection contrasts strikingly with the ways of handling garbage in elite areas serviced by private companies like Bin It, where residents are not seen handling their waste in public, simply leaving it out—without being accused of dumping—for an invisible other person to take away. By contrast, the

Figure 5. Domestic rubbish left for loaders on self-loading collection day in Katanga.
Photo by Jacob Doherty, 2014.

intense physicality of the self-loading mode of collecting waste is highly visible and collective.

Once the truck was full, Kato and I jumped back into the cabin. I expected us to head north to Kiteezi Landfill, a fifteen-kilometer drive through the congested center of town. Instead we turned south, through prosperous suburbs, past Speke Resort (Kampala's fanciest hotel, where foreign dignitaries typically stay), and, finally, down a dirt road that led us to another para-site: the marshy edges of Lake Victoria. The driver explained that they saved fuel by coming here instead of going all the way to the landfill. The landowner actually wanted them to dump here too, as the rubbish was helping to fill in the swamp, converting waste land into investment-primed lakefront private property. The truck emptied its load onto the soft, muddy ground, and the landowner and his young children began spreading it out and filling in the watery puddles. Kato told me that he knew that this deviation from the hermetically sealed waste stream went against the KCCA's policies. But it saved so much time that he could extract an extra truckload of trash out of Kabalagala every week, and so he feigned unknowing. Once empty, the truck headed back to Kabalagala to fill up three more times before, on its last circuit, stopping at the landfill.

Kato put in a lot of work all day, seeming to bring garbage collection to Kabalagala through sheer force of will. He mobilized residents; chose the route in coordination with the KCCA supervisors and drivers; and helped drag, lift, and toss garbage into the truck, never going more than a few minutes without answering his phone. He certainly could not have done this work alone, however, and relied on the labor of the Village Health Team, the CBO youth, and the municipal workers (fourteen people in total), not to mention the women and children of Kabalagala. Not every parish councillor dedicates themselves with such intensity to rubbish—Kato was in fact singled out and commended as the best councillor on waste management by the KCCA—and so not every neighborhood receives the same level of service as Kabalagala. Kampala's settlements are highly differentiated, in part due to the capacity and willingness of their leaders to extract services from the overstretched municipal government and to coordinate the huge amounts of labor required to sustain urban infrastructure like the municipal waste stream. Far from a homogenous "planet of slums," settlements are differentiated and differentiating, constituted by

significant differences in terms of infrastructural connection and municipal service provision that are, in part, effects of divergent levels of political
will and political connectedness.[11] Kato's membership in and loyalty to the
ruling NRM party, he admitted, was a useful resource despite the ostensibly apolitical and technical character of the KCCA.

Kato's intimate knowledge of Kabalagala's everyday disposal routines
and his careful choreography of municipal equipment, bureaucratic routines, and popular practices illustrates how infrastructure is assembled
in practice and offers a powerful contrast to the "sensitize and enforce"
paradigm that underpins approaches to waste management improvement
predicated on raising awareness. While for NGOs and municipal policy
makers sensitization operates on the assumption of an uninformed public
in need of awareness, local politicians like Kato use "the need for sensitization" as a way to resist or delay the implementation and enforcement
of unpopular or restrictive legislation that would negatively impact their
constituents, be it a ban on vending on the street, changing taxi routes,
or inspecting informal restaurant kitchens for sanitary conditions. In a
generic form, this stance could be expressed as "we are not opposed to
this change, which we recognize is part of development, but you cannot
enforce it until you have sensitized us about the new law." The temporal politics of delay here is an effort to forestall or even refuse unpopular
policies. As discussed in part 1, local elected officials' role becomes protecting constituents from development's displacements and the punitive
enforcement of modernization policies that are inadequately attentive to
the actual social and material relations of urban life.

9 Embodied Displacement

Cleaning is world making. It produces and sustains symbolic categories and social worlds. As Mary Douglas writes, "Eliminating [dirt] is not a negative movement, but a positive effort to organize the environment."[1] Like the social order that it, along with other urban infrastructures, subtends, the waste stream is no easy achievement; it takes work. This work is widely distributed in both space and time, extending from the everyday intimate practices of domestic waste disposal to the concentrated and mechanized weekly collections that punctuate waste management. Situated beside the formal waste stream, para-sites defer the hermetic sealing of the waste stream, simultaneously disrupting and enabling the flow of waste away. In their messiness, the waste stream and its para-sites illustrate the incomplete, multiple, and materially heterogeneous nature of urban infrastructure. Far from living disposable "wasted lives,"[2] salvagers, informal waste collectors, municipal loaders, and other inhabitants of Kampala's waste stream para-sites are engaged in the world-making work of cleaning the city. Why then does this work often appear to municipal planners as a form of pollution?

If for Douglas cleaning is world making, for philosopher Michel Serres polluting is world inhabiting. Serres argues that to clean is to prepare something for exchange. A hotel room, for instance, is cleaned to make it available to a new customer. Clean, the room has no rightful inhabitant and is freely available, at the disposal of an abstract consumer public. "Appropriation," on the other hand, "takes place through dirt."[3] Making place is a spatial process that unfolds in time, and it entails changing relations of possession: appropriation, occupation, and *taking place*. Inhabitation is the act of dirtying, of staining space with traces of one's body and life in such a way as to claim it as one's own: "Whoever spits in the soup keeps it."[4] Para-sites can thus be understood as polluting insofar as they bring together practices through which disposable people (in the eyes of official urban cleaners) inhabit and appropriate the city despite their disposability. Para-sites constitute a form of world-inhabiting direct action through which infrastructural services are not simply demanded but enacted.

Para-sites thus add complexity to the urban social and infrastructural order.[5] To deal with this complexity, to simplify the world, the KCCA has taken two courses of action: incorporation and expulsion. At Kiteezi Landfill, and with the variety of recycling kiosks within the city itself, the KCCA established relations of nonrecognition with salvagers, tacitly allowing them to break municipal ordinances and puncture the waste stream that official policy sought to seal. In other cities around the world, this accommodation has been taken even further as waste pickers have formed cooperatives, demanded political recognition, and have been awarded municipal contracts to partake in the formal waste management sector.[6] In Kabalagala and other low-income settlements in the city, on the other hand, the KCCA acted to expel para-sites, criminalizing informal waste collectors and arresting residents availing themselves of unauthorized dumpsites. In doing so, the KCCA was engaged in the world-making work of cleaning, defining, and enforcing aesthetic, environmental, and sanitary categories in ways that define and enforce specific norms and practices of citizenship and urban belonging.[7] Ecological and sanitary difference is thus used to justify and necessitate continued displacement and disposability. Frantz Fanon identified such dynamics as central to the production of space under settler colonialism. While the settlers' town, he wrote, is well built "and the garbage-cans swallow all the leavings, unseen, unknown and hardly thought about," conditions in the native town remain crowded, dirty, and unsanitary, laden with parasites that harbor the constant threat of contamination. This racialized ecological difference justifies and necessitates continued colonial displacement, "the abolition of one zone."[8] Vulnerability—exposure to toxicity and pollution—is translated into danger and the ontological nonbeing of disposability.

Away, too often, takes shape in bodies. Because they involve frequent exposure to the materiality of waste streams, para-sites entail hazardous externalities and debilitating subsidies, such as toxic body-burdens.[9] These moments of exposure are typical of what Rob Nixon calls slow violence, "calamities that patiently dispense their devastation while remaining outside our flickering attention spans."[10] Such exposures are part of the uneventful continuous noise of the everyday. These public health threats are what the normative waste stream is intended to guard against, concentrating Away in as few sites as possible to minimize risk. In Kampala, para-sites expose residents and workers to dioxins and particulate matter in the smoke of burning rubbish that can lead to cancer, asthma, and heart disease, lead and other heavy metals in water sources near dumping sites, and *Helicobacter pylori* (*H. pylori*), a bacteria commonly found in organic waste that can cause ulcers and stomach cancer.[11] Mohammed, the cofounder of the Waste Pickers Alliance referred me to the latter, a bacteria that he suggested affects most salvagers and loaders in the city but that, because medical staff at the local clinics most frequented by these workers are generally unaware of it, remains untreated, causing long-term damage. While researchers have documented the prevalence of *H. pylori* among children living with HIV in Kampala, there has been no research to date on its impact on waste workers in Uganda.[12] Studies in Mexico found *H. pylori* in two-thirds of landfill salvagers there.[13] Despite the social, medical, and economic tolls these kinds of exposures take, they do not attract the same levels of care or attention as more eventful disasters.

In keeping with Julie Livingston's observation that cancer remains largely off the public health agenda in Africa, where epidemics and outbreaks have occupied the medical imagination at the expense of longer-term forms of injury, I rarely heard mention of these longer-term dangers.[14] This is true more broadly in public health's framings of municipal solid waste management. These focus on diseases such as cholera, typhoid, and malaria but not on cancers, heart diseases, or respiratory illnesses contracted from long-term exposure to, for instance, the toxic smoke produced by burning rubbish. Despite the massive amount of medical and global health research generated in Uganda I located few studies on the long-term carcinogenic effects of waste management in the city.[15] Salvagers, managers, and activists alike focused on shorter term risks of injury and epidemic. The landfill engineer, for instance, compared salvagers to residents of fishing villages insofar as the specific conditions of their work (nonproductive gathering of salable materials) make them particularly vulnerable to HIV. While this association underpins much HIV/AIDS policy and research regarding fishing communities in Uganda, I have not found any research suggesting salvagers are more or less likely than the general population to contract HIV, and salvagers I interviewed did not mention this among their primary health concerns.

Salvagers and informal waste collectors nonetheless immerse themselves in the hazards of the para-site. Their understanding of these risks, and their willingness to take them on, speaks to the multiple intersecting forms of slow and structural violence—economic, environmental, and medical—that constitute para-sites and para-sitic encounters with waste. These forms of injury and slow death are thus multiply infrastructural. The ecology of para-sites magnifies other forms of structural violence and social inequality, channeling injurious flows into the bodies and environments of the urban poor. Yet they remain largely invisible. Like infrared waves, they are below the narrow spectrum of registerable harm produced by the sign of crisis that hangs over Africa, and the resultant temporal foreshortening renders long-term and complex injuries like cancer virtually unimaginable.[16] As Rodgers and O'Neill argue, infrastructure can be both a "material embodiment of violence (structural or otherwise)" and "its instrumental medium."[17] Attending to the slow violence that accrues in infrastructures' diverse para-sites suggests that violence can be, itself, infrastructural: a constitutive and radically unequal distribution of injury and vulnerability that is naturalized and obscured at the foundations of urban life.

In the context of ongoing displacement, however, vulnerability to infrastructural violence is not translated into a need for social protection but into another reason for disposability, as the poor are framed not so much as vulnerable but as always-already diseased, contagious, and in need of containment or removal. Despite the valued services they provide in Kampala's low-income neighborhoods, para-sites are rendered disposable and displaced. They do not, however, go Away.

PART III   Racializing Disposability

10   From Natives to Locals

Under the slogan "waste to wealth," NGOs in Uganda propose a novel solution to Kampala's ongoing garbage problem: make people see the latent usefulness of garbage and they will stop disposing of it carelessly and start gathering, sorting, and reusing it. Making biomass briquettes emerged as the most widespread project undertake by Community Based Organizations (CBOs) to turn waste to wealth. An alternate cooking fuel to replace charcoal and wood, biomass briquettes are made using the high-carbon organic wastes collected from the homes in a CBO's immediate neighborhood. Harnessing the population's entrepreneurial energies, briquette-making projects promised to tackle the city's environmental and economic issues simultaneously, bringing income and cleanliness to informal settlements like Bwaise, all while combating deforestation. The process was relatively easy to learn and, in its most basic form, required little capital, so it spread quickly as a staple activity of many CBOs interested in promoting hygiene, sanitation, and livelihood campaigns in low-income areas. Bwaise Tusobola (We Can) Development Association (BWATUDA), a small CBO based in the flood-prone Bwaise neighborhood with about thirty members including a core team of five leaders, was one such organization. When I first met them in 2013, BWATUDA had been making biomass briquettes for over a year as part of an ongoing effort to manage waste, improve sanitation, reduce flooding, and create sustainable incomes.

BWATUDA's neighborhood, Bwaise, appeals to low-income residents because of its low rents and because its relative proximity to major commercial centers means that transportation costs are affordable. Built on low-lying land tenuously reclaimed from Lubigi Swamp, the unplanned settlement has inadequate drainage and sanitation infrastructure and is prone to regular extreme flooding. In addition to its floods, Bwaise is represented in the media as a congested and overcrowded area with a reputation for poverty, crime, drug use, prostitution, and HIV.[1] In this rhetorical construction of the slum, residents themselves are held responsible for the fact that their settlement is unplanned.

Garbage plays an important part not only in this moral representation of "undeserving" subjects but also materially in making Bwaise flood. Uncollected garbage washes into the area's open drains, causing blockages that force storm water out of the drains and into homes and businesses. Floods are moral events. Widely framed as evidence of the problematic character of Bwaise's residents rather than decades of infrastructural neglect, they consolidate the neighborhood's place in the urban imaginary as part of the "bad city."[2] The KCCA individualizes the problem through the moralized figure of the litterer, arguing that if residents ended the vice of littering, then drains would not overflow. In conversation, however, KCCA officials recognized that the accumulation of garbage in drains is an inevitable consequence of the low rates of collection they provide combined with poor residents' inability to pay for private waste collection. In this context, CBOs like BWATUDA are touted as the solution to Bwaise's infrastructural problems. Through CBOs, communities are "empowered" to solve their own problems, charged with self-provisioning municipal services and improving their moral and material character.

Morally stigmatizing the urban poor, these present-day interpretations of the material conditions of the neighborhood as evidence of the immorality of its residents are a continuation of racist colonial conventions representing African bodies as filthy and polluting dangers that threaten the establishment of a reputable modern urban order.[3] Outside of South Africa, contemporary urban Africa is rarely studied through the lens of race. As Jemima Pierre has argued in her vital work reopening the discussion of the racialization of African urban space, there is a common assumption that, after independence, class and ethnicity replaced race as the dominant socio-spatial organizing logics of African cities.[4] This idea is evident, for example, in novelist Chimamanda Adichie's assertion that growing up, "identity in Nigeria [was] ethnic, religious . . . but race just wasn't present."[5] Adichie's point echoes Zora Neale Hurston's famous observation that "I feel most colored when I am thrown against a sharp white background."[6]

In postcolonial African cities, because the sharp white background of colonial rule is no longer apparent, there has seemed to be less imperative to analyze processes of racialization. In Kampala, anti-Black racism occasionally becomes an object of public critique and discussion around limited discrete events where discrimination is extremely explicit, such as a 2018 scandal involving a hotel advertising racially ring-fenced job opportunities for a white hotel manager and an Asian restaurant manager and a 2014 controversy in which a Ugandan journalist was labeled a sex worker and denied entry to a popular nightclub. The underlying structural processes of racialization remain largely absent from public discourse and academic attention. The problem, Pierre shows, is that this inattention to race obscures the ways postcolonial politics are formed by the transnational historical processes of extraction, accumulation, and (de)valuation of racial capitalism.[7] Pierre describes the way Ghana's independent government inherited a racial state built on the racial logic of indirect rule in which a white ruling class was structurally opposed to a native administration. This structure, she argues, obscured the racial nature of colonial rule insofar as most African encounters with the colonial state were with "traditional" or "tribal" authorities who governed "native" subjects according to their "customary law." This structure stabilized white rule by emphasizing ethnic difference rather than the commonality of racial domination. Pierre argues that Nkrumah's and subsequent Ghanaian administrations largely left this structure—along with the dominant position occupied by European firms in the commercial economy—intact, ruling via ethnic categories even as they articulated the racial discourse of pan-Africanism.

Today, racialized rule is most apparent in the institutional structures of the international development industry. The following chapters focus on the ways that encounters between Ugandans are mediated by the racialized scripts and meanings of development. To contextualize those encounters, this chapter situates them within the transnational racial formation of the development industry and its constructions of race and scale. Racialization takes place not only through the differential mapping of people onto space and into time—developed and developing, advanced and backward—but also through their projection onto scales.[8] Part of the coloniality of the neoliberal development industry is the way its mapping of global and local reproduces the structures of indirect rule: white rule becomes global institutions, natives become locals, and tribes become communities. This scalar hierarchy from the local community to the cosmopolitan global reproduces long-standing racial hierarchies while working to obscure them from everyday urban experience. It simultaneously constructs a set of upward aspirations for organizations like BWATUDA and other Ugandan participants in the development industry and limits the possibilities for realizing these ambitions. Funding, power, and the recognition of expertise are unevenly distributed through this hierarchy, in which the role of community-level actors is often to authenticate narratives authored elsewhere by producing images, stories, performances, and encounters as discursive resources to be mined and exported.[9] While community is a moral framework through which these performances are valorized, it also confines the performers to its racialized scale. Urban garbage has become entangled in these scalar hierarchies as a problematic and as a substance that CBOs in Kampala can mobilize to appeal to more powerful development industry actors. Waste worlds delimit the place from which CBOs engage with a range of NGOs and government bodies, and can participate in development discourses that elicit particular performances of cleaning and recycling.

Toward the end of a cleaning exercise organized by the KCCA one Saturday morning, once all the drains had been cleared and the volunteers had little left to do, a flood of paperwork rushed through the group gathered at the edge of an informal settlement. People clustered in groups, initially based on existing organizations, but then circulated from group to group. Someone handed me a paper, an activity log with the name of a CBO at the top. They wanted me to sign in to the cleanup exercise, giving my name, number, email, and affiliation. As soon as I'd filled out and handed back the form, I was given another one, from another organization, then another, and another. Fourteen in all. I found a more secluded shaded area and asked a volunteer what was going on. These organizations all needed to verify their activities in order to have a chance to secure more municipal funding or support from the myriad international NGOs based in Kampala. To verify themselves they needed names on paper, names that would later attest to their ability to do community outreach.

Notions of community are located at the center of the project of urban developmental respectability. The idea of community is essential to the ways in which urban poverty is racialized, temporally othered, and tethered to particular scales. NGO and government rhetoric uses *community* as a synonym for *slum*, indicating a bounded spatial entity, its inhabitants, and the presumed modes of relationality that they share. Community operates as a value that is vital to the infrastructure of feeling, but it must be materialized and rendered bureaucratically legible. Volunteers lament its absence as they clean—as in the rhetorical question, "Where is the community?" posed in chapter 14—but they are required by the forms of accountability and transparency demanded by the municipal development apparatus to document their engagement with the community. In order to fulfill these requirements, volunteers produce community on paper, materializing it in the form of signatures on sign-in sheets. Community names both the agent and object of development: the nexus of "backward" beliefs and behaviors as well as "primitive" material conditions and infrastructural services that need to be uplifted and civilized. Community thus appears as something already existing out in the world that volunteers and municipalities can go out into to do good works. As such, community is both seen as the potential answer to infrastructural abandonment (getting communities to clean themselves) but also fundamentally lacking in something that volunteers can bring (the *feeling* of responsibility).

As the role of the state has been reimagined and redirected in the decades since structural adjustment, communities are figured as both the beneficiaries of development, and, increasingly its agent.[10] The rise of the NGO as a global governing force is owed in part to "the assumption that they are less hierarchical, more democratic, more devoted to welfare and to serving subordinate or minority populations, and more cost effective than states. . . . NGOs themselves often rely on discourses of connection to grassroots movements."[11] CBOs play a vital role in this discourse.

When they partner with international NGOs to translate agendas into projects, to recruit participants and organize activities, to supply photo opportunities and narrative evidence, and to interface with local politicians and other leaders, they materialize the connections constitutive of a transnational NGO economy.[12] As bureaucratic representatives of community, CBOs are NGOs' preferred partners, authenticating globally circulating best-practice projects by providing evidence of local participation and consultation.[13]

This NGO economy assumes the existence of communities as spatially bounded entities with coherent shared interests and capacity for collective action among residents, positioning community leaders as the transparent mediums of these interests and capacities. In this way, communities play a critical role in neoliberal reforms toward cost recovery, appropriate technology, and decentralized service delivery. Community is gendered in such a way that the work involved in this kind of infrastructure is naturalized as an extension of domestic labor.[14] Based on research in Dakar, for instance, Fredericks shows how community becomes a technology of governance through which women's unpaid labor collecting trash, sweeping streets, and policing their neighbors is recruited as urban infrastructure and in turn is naturalized as a traditional part of women's role caring for and cleaning the home.[15] My research with a CBO working to transform domestic garbage into a new source of biofuel also made apparent the ways that, in addition to constituting a gendered division of labor, in the context of African cities governed through transnational "rule by NGO," community is also a highly racialized and racializing scalar framework.[16]

Positioned as both aid recipients and aspiring aid providers, BWATUDA's members shared a real desire to improve the material conditions of life in their neighborhood, combining desires to earn a living, establish professional identities, and achieve socially valued roles as community leaders. BWATUDA's founding members all had other jobs, working as a schoolteacher, motorcycle taxi driver, shopkeeper, porter and handyman, and driver for a safari company. But these jobs did not make them properly developmental subjects, due to a discursive emphasis on entrepreneurialism that privileges job creators and devalues wage earners as dependents. BWATUDA energetically took up the requirement to demonstrate its own "techno-moral" virtue: its capacity to ameliorate the moral and material condition of the community in which it is based.[17] BWATUDA, they hoped, could begin as a CBO and become an NGO in its own right, giving them a seat at a bigger table and access to a larger slice of what was often referred to as "the development cake." This scaling up was a well-established but hard-to-navigate path for small organizations.

Waste became an important substance upon which BWATUDA could perform a form of labor that would move them up the development industry hierarchy. By immersing themselves in the rotting banana peels, handling soot-covered equipment, shoveling mounds of char, and convincing neighbors to get their own hands dirty sorting waste, BWATUDA sought to render themselves socially valuable and recognizable as developmental entrepreneurs. In order to make briquette production viable, however, BWATUDA needed the financial support of national or international NGOs, which in turn required rendering themselves legible and legitimate in the eyes of funders. Their legibility and legitimacy were produced through performances of cleaning and recycling for international audiences who frequently toured this easily accessible geography of poverty.

One overcast morning I joined two of the founders of BWATUDA, Ismail Ntale and Daudi Lukonge, to make a big batch of briquettes. Ismail and I laid out a tarp and covered it with fresh peels, waiting for the sun to come out and dry them. Daudi lit the charring drum, emptied out a sack of already charred peels onto another tarp, and made some phone calls. Soon, seven other BWATUDA members joined us. Three women began crushing char while one of the other founders, Rose Birungi, boiled water and went to fetch cassava powder. Soon we had a tarp full of paste and Daudi and three other young men began filling up the briquette presses, pounding the paste into shape, finishing the tops, and setting the completed briquettes out to dry. Ismail's phone rang, and he announced to the group, "OK, they're coming!" He went to change into a fresh shirt. Along with most of the other women in the group, Rose discreetly walked away. Soon, a small tour bus pulled up on the nearby street, and twenty people descended.

The tour group, comprised of East African, European, and Indian NGO staff, gathered around to watch the production process, take photographs, and comment on the conditions they observed. "Look at these puddles! Can you imagine?" one visitor declared. A videographer set up a tripod near the charring drum and pointed his camera at Ismail, who was welcoming the visitors to Bwaise, introducing BWATUDA, and explaining their projects. Ismail emphasized the challenges that people in Bwaise face, such as poor sanitation, flooding, disease, and unemployment, which means "everyone stays at home" or "engages in sex for money." Here, as in media representations of Bwaise, waste, flooding, and commercial sex intermingle as evidence of the area's conditions.

BWATUDA was trying to tackle these problems, Ismail told the tour, by turning flood-causing trash into an employment opportunity that can reform abject environmental and moral conditions. Rose told me later that she and the other women in the group had disappeared from the scene because they did not want to be seen "dressed like this," referring to the comfortable but informal wraps and T-shirts she had spent the morning working in. She said that it was uncomfortable to "have those people looking at us here," gesturing to the puddles, the smelly latrine, and the mangy dog that often lingered around. The visit made Rose embarrassed, the influx of professionally dressed NGO staff highlighting the impoverished conditions of her own life. Without time to change into the nicer clothes she wore while teaching, she felt insufficiently distinguished from her surroundings and unwilling to be presented as part of the morally suspect slum that Ismail described.

After Ismail had set the scene, Daudi stepped up to go over the production process. He quickly switched into his more comfortable Luganda, and a tour leader tried as best as she could to keep up with translating his rapid-fire explanation. The group began to lose focus and talk among themselves. "Why aren't we supporting such projects?" asked a Ugandan woman. "This would be so cheap to start. They don't need much." "Yes!" an enthusiastic British colleague replied, "It's great because it is all on their own initiative. We could be doing this in Tanzania and Ethiopia too!" The group crowded around to watch Daudi demonstrate the penultimate step of the production process, pushing briquettes out of the press. He handed around nearly dried samples for the study tour to handle and shared a visitors' book for everyone to sign before the group's leader ushered them back into the bus. The international visitors were suitably impressed with Daudi and Ismail's performance but expressed their pleasure through a vision in which *they* would be the ones to make this project mobile and regional, skipping the

Figure 6. BWATUDA's charring drum in use in Bwaise. Photo by Jacob Doherty, 2013.

targeted question of funding *this* group by imagining projects at another scale. Later, Ismail explained that these people were from two NGOs "doing water." They had come to visit BWATUDA before, but they did not offer any support: "They just want to make it look like they are involved in so many communities, but when we call them, they cannot answer."

Despite the rhetoric of waste to wealth, briquette production was not producing much income for the group. The small revenues they earned selling briquettes to neighbors did not compensate for the labor-intensive work of collecting, sorting, and processing waste and then producing the briquettes. Ismail, Rose, and Daudi struggled to keep the other members interested. "Why would you come and work freely," Rose asked, "when you have people to feed at home?" Without regular clients, they had little incentive to produce regularly, and without more machines they could not scale up their production to the level required to attract more institutional customers with a high and constant demand for cooking fuel. Rose was of the opinion that the government and NGOs should be doing more to pro-mote and create a market for briquettes. "People don't know this product," she explained, "and so we can make it, but it is hard for us to get that mar-ket." NGOs proved very capable of promoting waste-to-wealth programs to CBOs, but there was no broader effort to advertise the product to con-sumers, and marketing was left up to each of the small, underfunded, and untrained community organizations to take on individually. The largest briquette producers in Uganda at the time of my research were an inter-nationally funded eco-business operating a large-scale production facility in a neighboring town. They hired dozens of young promoters to stand outside city supermarkets to market their expertly branded and packaged briquettes to the city's middle class and familiarize them with this all-new product. Even with this effort, they struggled to carve out a niche.

Daudi and Ismail's proposed solution to these problems was to scale up. To actually make enough money to compensate members for their work, BWATUDA would need to find a designated plot of land, acquire safety equipment, get more charring drums, expand their capacity to dry peelings and finished briquettes, secure machines that would cut down on the arduous work of pulverizing char and pounding briquettes, buy more presses, and train more members in production. To scale up the needed funding, they had to appeal to donors, the amorphous category of ever-present yet elusive development organizations that saturate Kam-pala. To raise funds for these expenses, Ismail—a job seeker with a uni-versity degree in social work—began to craft a concept paper, a written proposal outlining BWATUDA's aims, projects, environmental and social impacts, and its costs and benefits. This and other documents emphasized the moral virtue of the CBO, describing BWATUDA as a transformative force that would clean the neighborhood, inspire neighbors to engage in development, and bring hope to a benighted part of Kampala. Enrolling my help in this process, Daudi met Ismail and me at a nearby internet café to work, bringing drafts of the proposal so he could confirm the technical details of the production process.

Eventually, after several long meetings with the rest of the association and an extensive search to find an internet café in Bwaise with both a stable power connection and a functioning printer, the proposal was ready and the cover letter signed. Dressed up in jackets and ties, Ismail and I crossed town by motorbike to the southern suburban neighborhood where NGOs proliferated in large homes converted to office space. At the first NGO, a receptionist greeted us cautiously and listened as Ismail explained that we would like to deliver a concept to the project manager. The appropriate person was out of the office, "in the field for a study tour," the receptionist explained, but she would give him our proposal when he returned. This scene was repeated at the next NGO, where everyone in the office was out attending a workshop on public-private partnerships.

The experience of finding the NGOs' offices empty left Ismail dejected, feeling the highly unequal terms of connection between his organization and those he sought to solicit. He registered the vast chasm between his world and the moneyed world of NGOs, a world in which the production and consumption of scenes like the one he had just staged authenticates and lubricates the flow of wealth in the form of donor funds. While BWATUDA and its members had been available for these NGOs' visits, elaborately staging their community work in the hopes of accessing donor funds, NGO staff remained elusive. Ismail has to struggle to get across Kampala to deliver a proposal, but the NGO staff's mobility, the ease with which they could visit BWATUDA and casually propose "doing this in Tanzania or Ethiopia," kept them out of reach. This radical difference in mobility that Ismail encountered is one of the mundane ways in which NGOs are spatialized as "above" CBOs, able to drop in on BWATUDA for study visits that cannot be reciprocated, creating an affective experiential sense of "vertical encompassment" embedded in everyday metaphors and mundane bureaucratic practices.[18] The performance reinforced BWATUDA's position below, as a merely local entity with limited possibility and importance.

Such scalar encompassment is central to the construction and meaning of racial difference in the development industry. This resonates with the way the term *local* is used more broadly in Kampala to denote second-rate or low-quality goods, places, and even people that fail to meet the standards imputed to the foreign and the international. In addition to the racial theory that underpins much of the discourse of development, the practices, institutions, rituals, and everyday operations of development— including the representations of recipients, the patterns of sociality of development workers, the prevalence of English as the lingua franca, the rhetorical distinction between migrants and expatriates, and the location of headquarters in the metropolitan centers of the global north—reproduce racial hierarchy and global white supremacy.[19] In addition to the different pay grades of local and international staff, African development workers well into their careers are often subordinated as representatives of local interests and liaisons with local communities; are assumed to have limited scalar knowledge, capacity, or training; and are thus confined to particular scales and geographies of authority and mobility.[20] Localness is contrasted with the supposed universality attributed to whiteness in the development industry.[21] In this way, the local takes the place of the native, as constructed through the racial structure of colonial indirect rule.[22] Concepts like the local and the community are thus part of what Pierre calls the "racial vernacular of development," the conceptual architecture that scripts and scales development narratives and institutions and circumscribes the differential roles racialized subjects play in that project based on presumptions of African inferiority and lack.[23] Rather than a broad racial category of natives inhabited by a variety of specific tribes or ethnicities, contemporary development discourse is predicated on the even more obscured but nonetheless racialized categories of the local, populated by fuzzily defined but tacitly morally benevolent communities.

Activists like Daudi, Rose, and Ismail find themselves in an ambivalent position in a hierarchy shaped by the norms and expectations of community-based development. They have taken up the mantles of self-responsibility, participatory development, appropriate technology, and low-cost service delivery, but they remain dependent on external funds to actually meet many of these responsibilities. Community development has structured a niche in which they can act on their immediate environment and seek recognition, and it has simultaneously confined them to that niche and the forms of mobility, agency, and capacity for transformation that it entails. The local here operates as a racialized scalar trap that structures action while strictly delimiting its horizon.

11 Infrastructures of Feeling

In December 2013, the KCCA posted an image online announcing the unveiling of plans for a new park in the heart of downtown. The image is bland, a digital architectural rendering of urban space. It shows a green lawn with winding footpaths and young trees against the backdrop of high-rise apartment buildings in glass and stone. Currently fenced off as the private grounds of the Sheraton Hotel, this land has been reimagined as a public park and green space for the urban population. Yet, as a commenter points out, the future of clean and green public space in the city is populated by *bazungu* (white foreigners). They stroll, bike through the park, and picnic in the shade of the trees. The following chapters take up this commenter's question: "Why are there only bazungu in the 'new gardens'?" This image should be interpreted not merely as a telling lapse in architectural rendering; rather, it discloses an uncomfortable and underrepresented facet of urban environmental politics in Uganda, and throughout Africa—the implicit racial imaginary of urban transformation. The visible whiteness of this image of an urban future and its erasure of Black life, while jarring, is consistent with the racialized and racializing norms, expectations, and orientations that structure city-making projects.

Colonial cities were "racial projects": machines for hierarchizing and regimenting racialized bodies, assigning meaning to racial difference through the production of space.[1] If colonial urbanism not only reflected racial categories but also actively produced racialized subjects and populations through their built environments, it would be a mistake to assume that independence entirely interrupted the processes of racialization that had been built into cities' infrastructures and environments. Urban infrastructures like cordons sanitaires, drainage networks, and piped water systems were constructed not only to privilege white elites but also to maintain and manage the meanings of racial difference and spatialize differences in bio-political governance. These infrastructures remain in place in Kampala, forming the grid upon which postcolonial urban development has taken place.[2] Neocolonial racial divisions of labor stretch across national boundaries through the elaborate logistics of off-shoring, for example. White-Black hierarchies thus remain structurally in place if phenomenologically occluded. While political sovereignty has changed hands, the imperial racial project has remained partially in place, changed but not eradicated by anticolonial nationalism. Neither independence, broadly achieved in the 1960s, nor the "democratization" of the 1990s dismantled the colonial extractive economies that defined Africa's place in the world. While postcolonial governments "could no longer be challenged in explicitly racial terms," Pierre argues, "certain state apparatuses still reflected white European foreign economic and political domination."[3] In this context, as Kimari and Ernstson argue, infrastructures reproduce "a permanent subtext of racialisation across centuries" that is predicated on notions of African lack that reframe foreign investments in extraction as a humanitarian gift, while institutionalizing racial divisions of labor in project conception, planning, construction, and management.[4]

One aspect of this historical racial project has been to consolidate an idea of the African subject as an inherently rural person for whom city life represents a major cultural transformation. Respectability has been a central moral category through which this transformation is achieved, evaluated, and regulated. These meanings and morals, central to the racial project of colonial and postcolonial urbanism, shape the way contemporary Kampalans encounter waste, and encounter one another. To begin to understand the presence and effects of racial projects in contemporary Kampala's urban infrastructures, this chapter outlines the concept of infrastructures of feeling in affectively charged public cleaning events that took place throughout the city between 2011 and 2014. Despite the apparent absence of visible racial antagonism, these campaigns are nonetheless rich sites in which to examine urban transformation as a racialized, and racializing, process.

High-profile public cleaning events organized in the city by both the municipal government and NGOs perform the process of urban transformation in micro. Targeting discrete bounded areas; mobilizing the local community to gather garbage, sweep streets, and unclog drains; and sensitizing them to the issue of waste, these public spectacles ritualize urban transformation in a way that turns a long-term contentious process of spatial reform into discrete and uplifting events. Cleaning exercises rely upon and reproduce developmental respectability, an infrastructure of feeling that interpellates particular subjects as responsible citizens and agents of the city's transformation. In turn, it constructs other populations as the environmentally uninformed, unsanitary, and unruly others of the moral citizen. This infrastructure of feeling shapes how Kampalans apprehend their city and imagine transforming it, reproducing a racialized and racializing set of values and distinctions. Feelings about cleanliness, belonging, and futurity are central to the way that race takes place in postcolonial Kampala. Cleaning events organize and distribute affect in particular ways, turning encounters with filth into feel-good events. Why, and for whom, do cleaning exercises feel good? What is the infrastructure through which these feelings are produced and distributed? How are fraught and highly contested processes of urban transformation and displacement made to feel good? How do good feelings and affective investment reproduce social relations of abandonment?

Infrastructures of feeling are historically specific nexuses of morality, materiality, and ideology that configure emotional investment in particular futures, channel affective attachment, and distribute disposability. This idea builds on Raymond Williams's simultaneously suggestive and elusive concept of structure of feeling, referring to the ways "meanings and values . . . are actively lived and felt."[5] For Williams, a structure of feeling is not an epiphenomenal superstructure reflecting the real conditions of an economic base, rather it is a "deep and very wide possession in all actual communities precisely because it is on it that communication depends."[6] In this view, structures of feeling are themselves already infrastructural insofar as they are the basis—the shared dispositions and inhabited sets of sentiment, meaning, and value—upon which communication and thus culture, politics, and social life can take place. As Lindsey Green-Simms argues in her analysis of the postcolonial politics of urban mobility in West Africa, these public feelings, shaped in and around the built environment, structure both the routine interactions people have with cities' material infrastructure and their attachment to the state and its future-making projects.[7]

I use the term *infrastructure of feeling* in tandem with *affect* in order to theorize the socio-spatial relations, cultural logics, and materialities that mediate, distribute, and institutionalize feeling. While the concept of affect theorizes *intensities* that exceed fixed and bounded sociolinguistic categorization and exist in the intra-relations between bodies, spaces, images, and texts, infrastructures of feeling point to the ways in which affect is disciplined and harnessed to diverse sociopolitical ends.[8] The affects that infrastructures engender do not emerge sui generis. Cleaners' encounters with Kampala's waste reveal that infrastructures of feeling are vitally visceral, embodied, and tactile. These material and embodied registers are, nonetheless, infused with and inseparable from narratives of urban transformation, moralities of proper conduct, and histories of material investment in and abandonment of urban infrastructure. Infrastructures of feeling point to the ways in which affect is organized, channeled, and institutionalized in ways that sustain public-making projects.[9] Such public-making projects often focus affective investment in infrastructures themselves as bridges, roads, pipes, dams, and even dumps become objects of desire and attraction, harbingers of connectivity, progress, and modernity.[10] Beyond the general assertion that colonial legacies shape the postcolonial present by bringing together morality, materiality, and ideology, the idea of infrastructures of feeling offers a powerful framework for understanding the ways in which colonial racial epistemologies and hierarchies inhere in, and are remade through, processes of postcolonial urban transformation such as those performed in micro in public cleaning events.

One Tuesday morning in July 2011, I attended a community cleaning event organized by Green Life Uganda, the national branch of an international

NGO based in the United Kingdom. Green Life operates in three neighborhoods in Kampala, rotating this monthly event among them. This month they gathered to clean Kasubi Market, a busy roadside market at the heart of Uganda's most densely populated parish. The day began with a parade made up mostly of volunteer cleaners from Green Life and members of community groups funded by the NGO, led by a smartly uniformed marching band from a nearby secondary school, and escorted by an armed member of one of the city's myriad public security forces. Donning rubber gloves and wielding brooms and shovels, the group paraded back and forth along a six-hundred-meter stretch of a side street, calling passersby and local vendors, shopkeepers, and residents to join in. Few did, but many stopped to watch the parade with its bright blue-and-white banner informing them (in English) that "responsible citizens participate in cleaning and keeping their environment clean—you have the right to a clean and healthy environment" and instructing them, "*Wenyigire mu kuyonja ekitundu kyo*" (Join in cleaning your area).

The cleaning began slowly, with a few participants sweeping halfheartedly at the dusty litter along the sides of the roads. A few women wore brightly colored *gomesi*, complicated fabric wraps better suited for attending Parliament or weddings than for collecting rubbish. Others wore T-shirts and polo shirts bearing NGO logos and slogans promoting sanitation, HIV awareness, and the rights of the girl-child—evidence of participants' previous engagements with the city's dense development industry. Things got messy very quickly, however, once the shovels changed hands and a few of the women selling produce at the edge of the market packed up their stands, lifted the pallets on which they worked, and revealed the badly clogged drains below. Straddling these drains, a few young men and I started to shovel out the heavy silt, loading it into wheelbarrows pushed by eager children who, in turn, emptied them out onto the street. Bolder, the vendors stepped into the drains, shoveling quickly and without much apparent concern for the onlookers being sprayed with thick, muddy silt, plastic wrappers of all sorts, and rotting discarded food and peelings.

Twenty minutes into the cleaning, a faded green trash truck arrived from the municipality. A single worker started slowly and methodically transferring the heap of silt and rubbish from the street into the truck that would, in turn, take it to the municipal landfill at Kiteezi. The participants from the parade, by this point, were mostly standing back to watch. A man with a megaphone circulated through the crowd, thanking everyone for coming out to clean the area. Apart from a narrow strip, barely wider than a car, the street had been shut down for traffic, and an opportunistic hawker laid out a tarp to sell cheap plastic toys in the space between the trash heap and a car. Arrayed in this way, his mound of disposable goods was differentiated from the growing piles of garbage only through his work of framing: placing them on a tarp, calling out good deals to potential customers, and inviting people to inspect and buy his wares.

After an hour of shoveling in the muggy morning heat, I broke off to take some photos, jot down some notes (impossible to do in rubber gloves), and chat with Kenneth Lutaaya, one of the young men alongside whom I had been working. He lived in Kasubi, an area he jokingly, but somewhat proudly, called "my ghetto," referring to the area's poverty and associations with crime. By using the term *ghetto* (rather than the more commonly used *slum*) Kenneth was drawing on the lexicon of hip-hop in order to position Kasubi as one space among many in a global geography of Black masculinity, everyday hustle, and antagonistic countercultural production. This Black cosmopolitan vernacular offers a powerful alternative to the moralizing respectability politics that pathologizes poor and young Kampalans as idle, criminal, and without value.[11] Kenneth was trying to start a youth group to improve drainage in the more densely constructed residential areas inside the neighborhood. He had been attending these cleaning exercises every three months to try and secure a connection to Green Life and to get his youth group moving. Before I returned to Kampala in 2012, his project had come together, attracted funds from the municipality, and then fallen apart when all the members decided to divide and individually "eat" their collectively acquired funds. Kenneth lost his enthusiasm for youth activism and turned his attention toward his job in a downtown photo studio.

As he and I talked at the Green Life cleaning event, the man with the megaphone instructed us all to gather across the road in the small parking yard of a bar to hear speeches from local councillors, the mayor of the division, and representatives of Green Life. An MC urged the crowd to reflect on their environment: "Look around your place where you have put your items for sale. Is the place clean? Look and see where you have put the fish you are selling. Is the place clean? Look at where you have put the tomatoes. Is the place safe? In case the place is dirty, who is answerable to that place?" Having introduced the day's theme by shaming the vendors for the condition of their market, the MC brought on David Ndibalema, a program manager at Green Life, who thanked the audience for their attention and explained the organization's aim to "make sure that our market in Kasubi is the cleanest market. We would all want to come and buy food from here," he went on, "[but] I want to tell you, Mayor, that I live in Nansana [a suburban township located just beyond Kampala's boundaries popular with professionals], but many times I avoid buying food in Kasubi, and there are many reasons." He explained that this "market is one of the dirtiest in Uganda. But now we are coming to make it clean. It is your responsibility to tell your neighbor to keep your area where you are operating clean. And by doing that, you are getting more customers."

Next to speak was the mayor of Rubaga Division, Joyce Ssebuggwaawo, an up-and-coming politician known for her strong position on environmental issues. She thanked Green Life, rejoicing that people had come to help her voters in Kasubi with sanitation. Mr. Ndibalema's speech, she told her constituents, had affected her deeply: "David lives at Nansana, but he cannot buy anything from Kasubi Market because of poor sanitation. He further said that Kasubi is one of the dirtiest markets in Kampala, with very poor sanitation. This is too bad for us. I felt so embarrassed." Referring to the proximity of the market to Kasubi Tombs—one of the most important cultural sites in the Buganda Kingdom, where the four most recent kings are buried—she went on, "It is a shame that Ssaabasajja's [the king's] market, very close to the tombs, has poor sanitation to this extent."

Repeating the MC's plea, the mayor urged those present to "look around you. Look at where you are seated. How is the place? Is it clean? Look at where you are standing. Is the place clean? Look at all the shops we use. How are they? Are they clean? Let us collect the garbage in one place so that we are able to pick them. I am determined to ensure that we collect this garbage every day!" She ended her speech by encouraging "residents to ensure good sanitation" and explaining that "we are going to put up laws regarding sanitation. Whoever we find throwing garbage any how shall be dealt with in accordance with the law. We want you to be healthy."

Looking out from the stage where he was standing to address the crowd, Mr. Ndibalema could see the results of the morning's cleaning: a

municipal garbage truck parked on the side of the road at the busy intersection, a single worker tossing garbage up into the truck with a pitch-
fork, a massive and growing heap of drainage silt, trash, and market
debris marking the spot where cleaning had taken place. Mr. Ndibalema
had echoed the theme of the day, urging the market vendors to engage in
regular cleaning, to take responsibility for the conditions of their workplace. He ended his speech on an optimistic note: "I can see the market is
beginning to shine."

Green Life was just one of several governmental and nongovernmental bodies who regularly organized cleaning events like this in Kampala.
These affectively charged and highly performative events speak to multiple
publics, staging dramatic encounters with garbage in order to produce a
new urban future. Cleanups are performative in the sense of entailing a
fairly explicit degree of social performance and theatrical staging, as well
as in the sense of being constitutive—if not especially successfully—of
that which they name: "I dub thee responsible citizens." These speeches,
and the cleaning exercise that preceded them, manifest a particular infrastructure of feeling—a way of organizing affect to apprehend the present
and summon a possible future. Shame and pride, gratitude and embarrassment, rights and responsibilities, fear, disgust, and development were
all refracted through the staged encounter with waste. Rhetorically and
materially, cleaning exercises disclose a specific vision of the connections
and attachments that constitute the city. Fear of biological contamination
and the unruly intermingling of surfaces and substances keep the refined
NGO manager from coming to Kasubi, where he would otherwise be able
to support the people by buying their goods. In Mr. Ndibalema's rendering, markets (both as physical spaces and as the metaphoric means of
exchange) are the primary point of connection between the city's professional middle class and workers in the informal sector. Here, filth makes
evident the material and biological circuits running through the urban
population. Filth also gives rise to particular strategies of avoidance that,
in turn, block the circuits of consumption through which wealth would
otherwise, ostensibly, trickle down.

Speakers urged the vendors to use their senses in new ways. They were
instructed to look around them and see their market from an outsider's
perspective in order to fully comprehend the abject conditions in which

they work, to feel shame, and to respond by taking responsibility and materializing their constitutional rights to cleanliness.[12] In her response, as the political leader responsible for the area, the mayor performed embarrassment and shame on behalf of her constituents, suggesting that this should be their own emotional response to the outsider's gaze. She deployed ethnic attachments to compound this embarrassment, suggesting the proximity of the filthy market to the Baganda's Royal Tombs is particularly shameful. Mobilizing these feelings, she outlined an infrastructural fix, committing to securing daily municipal garbage collection across Kasubi and charging residents and vendors with the responsibility to gather their wastes together to enable municipal collection. Having inspired, informed, sensitized, cajoled, and threatened residents and market vendors, and having transferred the contents of a blocked drain onto the street, the cleaning exercise came to a close, heralding the beginning of a shining future for Kasubi and for Kampala, secured through new distributions of responsibility and sensibility and new moral-material relations with urban infrastructure.

Cleaning campaigns like this one enact a politics of respectability premised on the idea that abject urban areas and their residents must be redeemed and uplifted in order to prove their moral worth and substantiate their urban citizenship. They urge participants to remake themselves as moral citizens through direct, unmediated encounters with the city's most abject environmental spaces: clogged drains, uncollected heaps of waste, and the detritus of urban life accumulated in the city's margins. Through these events, other citizens are constructed as irresponsible litterers, guilty of environmental abandonment and unsanitary behavior, who are to be inspired and sensitized (if not policed and fined) to change their ways. This variegated process of subject formation has a particular temporal structure: visions of a bright and clean future ("I can see the market beginning to shine," as Mr. Ndibalema put it) produce distinct ways of sensing, feeling, and acting upon the present as especially abject, and of explaining the past that produced present conditions. These encounters are racialized and racializing, relying upon and reproducing notions of inherent and essential racial-cultural difference that haunt urbanization and urban transformation in Kampala.

As Kampala's cleaning exercises reveal, the material infrastructure of the city becomes an ideal and recurrent venue for channeling and recombining values and feelings. While I am concerned here with the sentiments that are invested in cleanliness, littering, and garbage collection, this infrastructure of feeling also takes shape around infrastructures of public transportation, electricity metering, water delivery, and the provision of sanitation.[13] As instantiations of urban transformation that render the amorphous process of city development into unique, ritualized events, the cleaning exercises organized by NGOs like Green Life Uganda and Kampala City Yange, as well as by the municipal government, are densely charged with the ideals, affects, assumptions, and anxieties that underpin contestations over Kampala's future. Far from purely local, this infrastructure of feeling has complex transnational historical lineages. Situating these cleaning exercises within the global histories of respectability politics reveals the vital ways that race centrally influences how Kampala's emergent class structure is lived, felt, and naturalized as a progressive facet of urban development.

12 Developmental Respectability

Development has taken the place of colonialism as the critical "power structure within which [Africanist anthropology] takes shape."[1] The development encounter precedes and facilitates but also frustrates ethnographic encounters. As a white foreigner in my late twenties at the time I conducted fieldwork, the vast majority of my interlocutors initially assumed I was in their city to "bring us development," as expressed by Kato Mubiru, the Kabalagala politician so dedicated to garbage removal that he proudly earned the nickname "Kasasiro" (Garbage). As Pierre points out, "wealth and aid continue to be intimately tied to race and whiteness," a fact that "allows racialized-as-white people and groups to possess status and power."[2] Such privilege was the condition of possibility for my ethnographic research: my passport granted me easy access to Uganda, my foreign university affiliation meant important policy makers took the time to talk with me, and whiteness meant my inquiries were treated not only seriously but as a potential harbinger of development and aid.

While ethnographic fieldwork sometimes transgressed the routine racial circuits of the city, bringing me into intimate contact with clogged drains and dumpsites as I sweated alongside other volunteers on cleanups, these transgressions tended to reinforce, not rupture, the structure of white privilege, becoming exceptional moments worthy of praise and gratitude for the humanitarian urge to uplift, which was assumed to motivate them. At the same time, the naturalized pedagogy of development as a white-to-black knowledge transfer frustrated my research insofar as it was often expected that I was coming to a place to teach people what to do rather than to learn from them about what they were already doing. I regularly asked environmental activists to tell me about the various recycling projects they had undertaken and was answered with self-deprecating comments about them and calls on my own assumed expertise: "You tell us what to do." Several times, after a few months of working with people in their neighborhoods, I was thanked (in a way that was so exceedingly effusive that it became uncomfortable) for "listening to us local people—mostly you whites don't want to hear us," in the words of Ethan Wekomba, a community health worker who had introduced me to many of his neighbors.

One afternoon in January 2014, I paid a visit to my friend Kenneth Lutaaya (whom I had first met at a cleanup in "his ghetto," Kasubi) at his downtown photography studio. He was hard at work airbrushing his clients' photos before printing them, but he was happy to take a break to chat. He put on a video he wanted to show me, Beyoncé's performance at the Grammys the previous weekend. He was stunned by the lighting and staging and wanted to talk about how he could emulate it in his studio. Once the video ended, Kenneth commented that many of his friends rejected Beyoncé because of her ties to the illuminati. Kenneth explained the prevalence of occult symbolism (especially triangles and the all-seeing Eye of Providence) in Beyoncé's videos, which, combined with her highly sexual performance served as evidence of links to an immoral New World Order that also included powerful figures like George W. Bush and Queen Elizabeth II. Kenneth wanted me, as an excessively educated person from the United States, to reassure him that this was not really true. I staggered through an improvised explanation of the difference between hegemony and conspiracy, and the fact that while the levers of global power may be located in a small number of institutions, they are operated in much more transparent and banal ways than the New World Order theories would have it. Kenneth interrupted, "Even if the New World Order is real, maybe it's better if you people, you whites, are in control.[3] What have we Africans done since independence? It would be better if you ruled us. At least then we wouldn't have this corruption and tribalism." Many anthropologists have documented their embarrassment by such articulations of white supremacy by formerly colonized people, but embarrassment is an inadequate intellectual response to the insight these encounters actually offer. Considered in light of the transnational construction of white supremacy and disposability that connects the histories of postcolonial developmental respectability and racial capitalist statecraft, these encounters disclose the foundationally racialized work of power, statecraft, political subjectivity, and the ways in which race takes place in and through the ideological, institutional, and affective infrastructures of development. Kenneth's expression of internalized subordination was as depressing and uncomfortable as it was familiar; rarer was an encounter a few days later when I was walking through Kisenyi, Kampala's oldest and most cosmopolitan slum, and an intoxicated young man made eye contact and yelled, "Fuck off white man!" At last, the anthropologist encountered the properly resistant subject!

When I did eventually "fuck off" back to the United States six months after the young stranger's urging, I returned to a country dramatically failing to deal with its own forms of Black disposability. In fact, the entire career of my research on disposability has been punctuated by Black disposability in the United States. I wrote grant applications during the fall of 2011 while joining a series of Occupy Oakland protests at the public square renamed Oscar Grant Plaza after the young Black man murdered by BART police officer Johannes Mehserle. During a break in fieldwork in summer 2013 to visit my partner in New York, I participated in a march through the city demonstrating against the verdict acquitting George Zimmerman in the murder of Black teenager Trayvon Martin. In the weeks following my return from fieldwork in 2014, in Staten Island, New York police officers Daniel Pantaleo and Justin Damico killed Eric Garner as he struggled to earn a living in urban America's unrecognized informal economy. Later in the summer of 2014, police officer Darren Wilson killed another Black teenager, Michael Brown Jr., in Ferguson, Missouri, sparking months of protests across the country and the emergence of #BlackLivesMatter as a political, cultural, and intellectual response to Black disposability in America. In 2016, police killed Alfred Olango, a Ugandan refugee in California, after his sister called to request psychiatric help handling his acute mental distress. As the final revisions on this book were being completed, in 2020 new waves of protests against police brutality, sparked by the murders of Breonna Taylor and George Floyd, continued to assert, in the face of the country's violent institutionalized anti-Blackness, that Black lives matter.[4]

These police killings, and the political responses to them, have reanimated arguments over the merits and limitations of the politics of respectability as a response to white supremacy. These debates and scholarship on the politics and genealogies of respectability offer salient critical insight for understanding the entanglement of uplift and abjection, of development and disposability, that characterizes the infrastructures of feeling through which Kampalans encounter waste and enact urban cleaning. In the United States, Black respectability politics emerged as a means of contesting American racism and of collectively navigating the transformation from slavery to freedom. As Susana Morris writes, respectability has been a "strategy for navigating a hostile society that characterizes Blackness as the definition of deviance."[5] It has always been contentious, lying at the heart of the split between radicals and reformers and recurring as a constant point of contention among generations of Black feminist intellectuals.[6] Multiple modes of racial uplift have been articulated over the years, from broad visions of collective aspiration and advancement drawing on liberation theologies to more narrowly defined elite projects seeking to qualify for liberal rights and citizenship through cultural assimilation.[7] Highly ambivalent, respectability has on the one hand offered Black Americans a means to confront and disprove egregious racist stereotypes, undermining the basis of post-slavery regimes of spatial and economic segregation. On the other hand, the vindicationist project allowed the dominant ideologies underpinning racial terror in the United States to set the terms through which Black progress would be imagined. Respectability in the United States thus has emerged as an articulation of racist and anti-racist discourse.

Across its multiple expressions, respectability politics has largely been predicated on "male dominance and Victorian ideals of sexual difference."[8] Racial unity was constructed through the rhetorical abjection and demonization of unfit and improper people, particularly of poor women who were excluded from the community produced through the politics of respectability.[9] Early twentieth century respectability enabled Black women, for instance, to construct positive self-images centered on the values of thrift, chastity, piety, cleanliness, hard work, and temperance, expressing these values through forms of dress, personal comportment, and manners, as well as through practices of homemaking and childrearing.[10] But, in creating this linkage between the care for the home and racial uplift, the politics of respectability held women uniquely responsible for the condition of the Black race.[11] Regardless of Black men's and women's efforts, these respectabilities have been no match for the racial terror and hierarchy of American society. Such fastidious forms of homemaking and self-presentation became a Sisyphean task, eroding the sense of self even as they provide a sense of social orientation and futurity.[12] By focusing on the private behavior of African Americans, the politics of respectability have privatized racial discrimination, positing nonconformity with the norms of respectable morals and hygiene as the root cause of racial injustice.[13] As such, respectability politics bear an uncanny resemblance to, and reproduce, the racist discourse it seeks to undermine.

In urban areas during the Great Migration, respectability was a means of managing the waves of African Americans leaving the rural South, protecting the fragile reputation of urban Black middle classes from contamination by southern migrants perceived as backward and primitive. Churches and other institutions sought to transform these migrants into good citizens, adjusted to the disciplines of urban life and labor. In doing so, however, they rigidified the distinctions of class difference within African American communities. Historian Kevin Gaines writes that in these urbanizing communities, "formulaic representations of urban pathology were crucial to [elites'] self-image of respectability," reflecting "a developmental construction of race and class that bestowed on 'better class' blacks an illusory sense of self-importance as it divested poor urban blacks of agency and humanity."[14] By embedding the norms of white bourgeois respectability in racial uplift ideology, reformers naturalized class hierarchy as the means of confronting racism and obscured structural critiques of racial order, labor conditions, and property relations with moral panics around individual pathological behavior.[15]

Public cleanliness has long been a primary element of respectability. Because the degraded conditions of Black slums were held responsible for "the immoral and even criminal behavior of their residents," Black leaders, in response, historically supported programs for slum clearance and beautification.[16] Anticipating 1990s "broken windows" policies, they hoped that by picking up garbage and painting houses they could remake the environment and in turn the moral life of Black urban communities. But in the context of racial capitalism and speculative urban real estate markets, projects to clean up neighborhoods often contribute to the devaluation of Black lives and spaces, naturalizing waves of displacement in the name of development in what anthropologist of urban wastes Marisa Solomon calls the "racialized temporality of betterment."[17] These forms of privatized uplift through personal cleanliness and comportment differentially distribute citizenship to some—exerting a significant chronic mental and physical toll—while relegating the majority of the population to spatial, political, and moral marginality.[18]

The politics of respectability continues to shape the moral imagination of anti-racism in the United States. Respectability rhetoric has allied civil rights leaders with the racially coded tough-on-crime rhetoric that makes up the public consensus underpinning the war on drugs and mass incarceration.[19] Respectability inserts a break within the population, dividing worthy citizens from a marginalized underclass unable or unwilling to conform to the moral norms of the middle class. As a strategy for liberation, Michelle Alexander argues, respectability politics is entirely inadequate: it too often becomes a form of victim blaming that obscures the structural relations of mass incarceration and, paradoxically, places further discursive emphasis on precisely the negative stereotypes that it seeks to overcome, authorizing further investment in the forms of surveillance and policing that fuel mass incarceration and "ultimately secure the hegemony of ruling social structures."[20] Rather than undermining white supremacy, the politics of respectability have entrenched a moralizing hierarchy of more and less killable subjects living more and less grievable lives, exemplified in utterances in response to police killings that victims are "no angel[s]."[21]

The point of recounting this history is not to argue that contemporary Africa is repeating a history already played out in the United States. Rather, it is to describe the complex and ambivalent motivations and effects of respectability politics in order to illustrate how projects of moral uplift can reproduce racist ideologies and reinforce public support for the institutions of disposability. Contestations around respectability politics make clear that racism manifests not only in violent moments of discrimination, displacement, and eviction but also in forms of care like cleaning exercises that both uplift and render disposable. Moreover, these are overlapping rather than purely parallel histories. African Americans have looked to the African past for models and traditions of Black respectability and power, an intellectual move that gives Afrocentric ideology its progressive edge versus white supremacy but naturalizes its patriarchal conservatism versus poor people and women.[22] Colonial states in West Africa turned to the institutions of Black American respectability, such as Booker T. Washington's Tuskegee Institute, to help forge racially appropriate systems of economic control and labor extraction.[23] Similarly, scholars of African family life have described coeval and interlinked histories of discipline and distinction that have shaped gender, class, race, and spatial relations in Africa. Respectability was a constituent element in the colonial civilizing mission to materially and spiritually transform African social worlds. This transformation relied on a "fastidious physical refashioning" of the human body and its habits, focusing on seemingly insignificant and banal details of everyday comportment.[24]

Across East and Southern Africa, British colonial pedagogies of respectability expanded outward from the body to the built environment, encompassing the application of soaps and lotions, modes of dress, food preparation and consumption, table manners, domestic architecture, care for the home, bed making, penmanship, and gardening. Accompanying these pedagogies of respectability were a suite of commodities and modes of consumption that enabled the cultural construction of new forms of urban subjectivity, class identity, and domestic life.[25] Colonial domesticities were fundamentally entwined with the medicalizing discourses of hygiene and sanitation and were predicated on understandings of African bodies as dirty and diseased. Hygiene thus emerged as an alibi for regimes of urban racial segregation. Combined with uneven infrastructural investment, segregation led to the uneven health outcomes and disease burdens across racialized populations that, in the circular logic of empire, justified segregation.[26] This medico-moral discourse shifted, as early as the 1930s, from a biological to a cultural explanation of racial difference, that, nonetheless, sustained the construction of the African as a radical other and object of medical-knowledge production.[27]

Respectability discourse was propagated by the often contradictory institutions of empire: schools, churches, missions, clinics, and the administrative offices of urban and economic planners.[28] In Uganda, as elsewhere, these institutions and their associated moralities became loci of class aspiration, such that upward mobility was inseparably linked to a racialized process of embourgeoisement. Colonial missionary education for elite girls had focused on hygiene, domesticity, and respectable conduct, institutionalizing the figure of the proper woman as the only means by which women could be liberated from "subordination inherent in their roles in traditional society."[29] Bourgeois domestic and gender norms were not uncritically or universally accepted, however. Aristocratic Baganda girls (or their chiefly fathers), expecting to inherit the privileges of idleness and sexual freedom associated with their position in the precolonial class structure, chaffed against missionary teachers' emphasis on agricultural work, domestic labor, and sexual propriety.[30] As Kyomuhendo and McIntosh argue, colonialism intensified and transformed extant gender roles in Uganda, narrowing the range of morally acceptable forms of womanhood.[31] As Ugandan women bore the weight and contradictions of transforming and modernizing the "traditional family," the intimate sphere of reproduction became thoroughly entwined with colonial and postcolonial statecraft.[32] This trend has only intensified in the era of AIDS where the links between respectability, moral cleanliness, sexuality, and (public) health have become even more affectively charged.[33] As in the United States, respectability has been a conduit for the white colonial gaze, working "to introduce bourgeois customs to poor Black women and to persuade whites of Black women's ability to adopt these customs."[34]

Historically, women's place in Kampala has been contested in the idiom of respectability, defined primarily in reference to sexual restraint and child bearing restricted to marriage.[35] For urban women, this has long entailed navigating the ambiguity between marriage and prostitution, vague categories encompassing a wide range of sexual and financial arrangements.[36] Respectability politics continues to mobilize shame to police women's place in the city, linking women to the protection of custom and family life in order to regulate economic life and public presence. The purported distinction between public and domestic space is, of course, radically undermined by the infrastructure and materiality of urban ways of life in low-income settlements where the domestic work of cooking, cleaning, and child-rearing takes place not in the privacy of the home but under the public scrutiny of the neighborhood gaze, an infrastructural nondistinction with important consequences for the gendering of respectability.

Contradictory forces have structured non-elite women's urban experience: economic needs push women into business activities and multipartner survival strategies, while a conservative moral framework emphasizes domestic respectability and care for the home and children above all. By becoming "proper women"—modestly dressed, living in clean homes, raising clean and well-behaved children, and partnered to a suitable man—many urban women have negotiated these stigmatizing discourses to secure their place in the city. Nonetheless, these agentive strategies have further marginalized single mothers, infertile women, and unsuccessful businesswomen, defining women's proper roles as those of wife and mother.[37] The figure of the "proper woman" has been frequently mobilized by elite women to disparage and marginalize low-income urban women. As Christine Obbo theorizes in her ethnographic study of a low-income Kampala neighborhood in the early 1970s, these gendered politics of respectability are also constitutively temporal, as women, charged as the bearers of traditional morality and custom, have been put in a position mediating between present and past, whereas their urban male counterparts, charged with the task of national development, mediate present and future.[38] Nonetheless, women have actively redefined and expanded the model of domestic virtue and have incorporated more and more forms of work outside the home and the family (initially expanding to take up care work as nurses and teachers, before consolidating more professional and entrepreneurial roles in both the formal and informal sectors).[39] The imprisonment of activist and anthropologist Stella Nyanzi in 2018 for publishing vulgar anti-regime poetry on Facebook illustrates the ways that respectability politics bolster authoritarian rule. Nyanzi's protests draw on the transgressive traditions of radical rudeness and "genital cursing" that women in Uganda and elsewhere in Africa have used to disrupt the bounds of respectability and denounce oppression.[40] In Nyanzi's transgressive radical rudeness, vulgarity is used to scandalize the public and draw attention to linked critiques of the criminalization of sexuality, the harassment of women street vendors, and the failure to deliver on the promises that brought Museveni to power thirty-four years earlier.[41]

Developmental respectability in contemporary Kampala is not simply a story of colonial imposition, however. The widespread resonance of this infrastructure of feeling owes to its entanglement with a variety of Ugandan moral logics and codes of behavior. The Buganda Kingdom was seen as a suitable agent of British indirect rule in the commonwealth not simply because of its territorial control and recognizable structure of authority but also because of the shared commitment to the idea of progress and because the perceived neatness and cleanliness of the kingdom's roads, gardens, and people made them ideal colonial subjects.[42] Ugandans have long adopted, adapted, and incorporated elements of missionary Christianity, colonial statecraft, and modernization theory toward their own ends and have engaged in lively debates about the proper moral grounding for development, the place of traditional law, ritual, and practices in a modernizing polity, and the diverse paths to collective and individual progress.[43]

One of the shared elements of these debates about development has been the metaphor of organic growth, comparing society to a child that is maturing to adulthood.[44] In *Do Not Go Back*, a 1938 manifesto against the return of Ganda customs that had been targeted by missionaries, for example, the Ganda palace intellectual Ham Mukasa opens with the claim that "the child, while he is yet young, all his customs remain young, but as he continues to grow, those customs of youth come to an end, because he is beginning to understand the good things and the bad things—to sense them himself and how the bad things bring him to grief, while the good things cause him happiness. . . . And so it is with all peoples in every nation."[45] The development of the nation, in this view, is akin to the progression to maturity such that there can be no going back. While this organic metaphor, common in Ganda political discourses, echoes agrarian cyclical ideas about growth and regeneration in ways that contrast with the mechanical linearity of modernization, it equally resonates with what political theorist Uday Mehta identifies as a fundamental strategy of exclusion constitutive of colonial rule: the categorization of colonial subjects as children in need of pedagogic instruction from European civilization in order to mature into the capacity for self-rule.[46] It makes little sense, however, to attempt to parse the indigenous from the imported elements of this infrastructure of feeling in its manifestations shaping encounters with waste and urban cleaning projects in the first decades of the twenty-first century, as these genealogies have become thoroughly intertwined. As in all modern postcolonial societies, multiple forms of racial ideology have become a constitutive element of the patterns of thought and structures of government in contemporary Uganda. Respectability in Uganda, as in the United States, has emerged not as the autonomous ideology of one population but as a set of ambivalent responses to changes in the structure and organization of racial hierarchy. The official end of colonial rule did not eradicate racist assumptions but refracted and reproduced them through articulations with new discourses, debates, and state projects.

Despite the nationalist rejection of colonial racial hierarchies, the structure of desire in which "Whiteness is at once development, modernity, intelligence, innovation, technology, cultural and aesthetic superiority, and economic and political domination" persists.[47] While American respectability politics emerged in the context of the racial terror of the Jim Crow South and racist northern cities and has evolved as an explicit critique of racist stereotypes, in Uganda the absence of anti-racist analysis in public culture and scholarship in the postcolonial period has meant that the racial politics of respectability have remained invisible and the universality of its bourgeois norms unchallenged. But, as Pierre observes, "race exists and is always expressed in excess of the corporeal," such that the reproduction of racial hierarchies can be discerned even beyond the embodied presence of whiteness.[48] This is particularly important because of the way that the transnational organization of the division of labor in the development industry distributes racialized bodies across national boundaries and gated enclaves. Linking American and African forms of disposability in this chapter is not intended to subordinate the political realities of the latter to the privileged analytic categories of the former. Rather, it draws on the critiques of the ways that respectability internalizes white supremacist racial ideologies in order to make visible the workings of racial logics in postcolonial Africa. My emphasis here has been less on the whiteness of imagined futures and developmental aspirations in

Kampala than on the ways that clean and green futures are articulated in terms of anti-Blackness. These connections signal the importance of the shared critical projects of pan-Africanism and the Black radical tradition that offer alternate visions of transformation. This shared transcontinental intellectual and political genealogy contains utopian dreams of freedom, futurity, and feeling that are far more radical than those contained within the confines of developmental respectability.

As performances that crystalize developmental respectability, cleaning exercises promise to remake the city, its space, and its inhabitants, but in doing so they reinscribe ruptures within the population, rhetorically constructing unequal access to infrastructural services as a form of cultural-temporal-racial difference. Casting behavior and culture as problems to be overcome, respectability politics is a form of cruel optimism, Lauren Berlant's term for forms of futurity and attachment that sustain and orient subjects while simultaneously reproducing attritional structures that erode, exhaust, and undermine them.[49] The highly individualizing moral infrastructure of feeling has become the basis of a developmental public consensus of uplift that obscures, while deepening, ongoing forms of racialization and marginalization. Developmental respectability codes *African, village,* and *local* as synonymous with *backwardness,* with a subordinate—childish at best and immoral at worst—condition that must be overcome in order to progress. Developmental respectability is thus an infra-structure of feeling that entangles shame and uplift, progress and embarrassment. It mediates the moral contradictions and disjunctive temporalities of urban development and social transformation, offering a means to make a place in the world through personal moral conduct, but in turn consolidating exclusionary hierarchies of race, class, and gender. These visions of the future also elicit particular revaluations of the city's and the nation's past. As chapter 13 shows, the form of developmental respectability these events manifest reflect a broader form of emplotment that shapes encounters with waste worlds. Problematizing present conditions in relation to the broken promises and missed opportunities of independence, this infrastructure of feeling draws on wells of both colonial nostalgia and precolonial, reinvented traditions.

13 Waste in Time

In September 2012 Kampala was in the midst of preparations for the celebration of the fiftieth anniversary of Uganda's independence. A new pavilion was under construction to expand seating at the parade grounds in Independence Park in anticipation of a daylong ceremony featuring military parades, performances of national dances, the awarding of medals, patriotic speeches, and a fly-by from the recently, and controversially, acquired MiG fighter jets.[1] The TV news dedicated a section to national history every night and the daily papers ran features remembering the lives and achievements of the country's national heroes and villains. Billboards around town congratulated the nation on its independence. Uganda@50 branding saturated the urban public sphere. A few times a day I would hear children at the school across the path from where I was living singing a rousing verse of the national anthem. The anthem was ever present, always on the radio, in the air, and stuck in my head, so much so that I often found myself humming "UGANDA! May God uphold thee. . . ."

Not everyone was entirely enthusiastic about the independence celebrations, however. At the time, I was staying with a family on the southern outskirts of Kampala. I asked the father of the family what his plans were for the celebration. Mr. Mulindwa, a retired civil servant in his seventies who remembered celebrating Uganda's first Independence Day in 1962 as a young man, replied, "What do we have to celebrate? How can we celebrate our independence when *this* is going on?" he continued, gesturing to the newspaper on the coffee table. The paper bore details of an emerging corruption scandal in the Office of the Prime Minister. Officials had embezzled US$13 million intended for post-conflict reconstruction and development in the north and northeast of Uganda.[2] In the wake of these revelations, the United Kingdom, Ireland, Germany, Norway, and Denmark suspended bilateral aid, and the continuation of World Bank funding also came under review. The scandal and ensuing aid freezes revealed some of the limits of Ugandan sovereignty insofar as nearly a quarter of the national budget was underwritten by foreign governments and multilateral donors. In the wake of these cuts, a raft of regressive taxes on water, paraffin, and other basic goods were introduced the following year to cover the ensuing shortfall by expanding the national domestic tax base.[3]

Despite the mood of national celebration in October 2012, Mr. Mulindwa was not alone in questioning the narrative of national progress and development. This sentiment was widely expressed in public writing in the national media. For many commentators, the condition of the capital city disclosed the true condition of the nation: dilapidated and in need of repair. Kampala's potholes, traffic jams, floods, open drains, and litter-strewn slums not only undercut the celebratory mood of commemorating independence but also, more profoundly, called into question the basic value and meaning of independence as such. In contrast to the positive waste of destructive creation, garbage in this view becomes an index of stalled progress and temporal failure. Writing in *The Observer*, editorialist Charles Odoobo Bichachi asserted "It is time we started taking ourselves a little more seriously, took charge of the problems of our country and do the basic things within our means to make the country run more smoothly, and make life a little better for our people, else the 50 years of our independence will remain meaningless."[4] What does independence mean, in other words, when surrounded by crumbling infrastructure? Without progress, what is the point of sovereignty, Bichachi wonders. Similarly, in a 2011 article marking NRM Day, a holiday celebrating the anniversary of the 1986 capture of Kampala by Yoweri Museveni's National Resistance Army, the year zero of contemporary Ugandan politics, political scientist and public intellectual Frederick Golooba-Mutebi noted how public filth reveals that "there has been something 'not quite right' with Uganda's postcolonial governments."[5] He gets this sense, he wrote in *The East African*, from a widely circulating chain email contrasting pictures of Kampala's clean and orderly colonial past with its filthy dilapidated present: "The colonists left clean, well-ordered cities and then we took them over and trashed them, even if many of us, like Museveni, keep our private homes shining."[6] Golooba-Mutebi blames a deficiency in national culture, a disjunction between citizens' care for the private sphere and abandonment of the public sphere, that has led to personal cleanliness but a filthy environment. In this view, responsibility for urban decay, manifest as public filth, is individualized as a failure of personal civic responsibility that, in turn, indicts Uganda's corrupted values and calls into question the propriety of its "not quite right" postcolonial governments. Urban infrastructure built "by the British" during the colonial period or by foreign governments' development programs have not been adequately maintained, it is argued, not because of decades of internationally mandated political priorities that have failed to address the needs of African cities but because Kampalans lack a sense of "ownership" over their city and have, in turn, failed to take responsibility and care for it.[7]

Echoing this theme, Bichachi asserts that "you can take the African out of the village, but you may never take the village out of him!"[8] He voices this sentiment as the thoughts of a hypothetical Japanese ambassador to Uganda, looking through a car window and being dismayed by the ways in which Japanese taxpayers' funds have been squandered on development projects in a city that cannot shed its *kavuyo* (chaos). Bichachi asks if the problem is that national infrastructure is foreign: "Is it because we never really built these things, that they were just thrown at us, that we do not appreciate their usefulness and therefore cannot devote a little effort on simple maintenance? Should the Queen of England return to fix the Owen Falls Dam Bridge?"[9] Rather than celebration, the fiftieth anniversary of independence thus provoked a national existential crisis that fixated on infrastructure and urban waste as indexes of an indelible backwardness that the postcolonial state had failed to overcome.

Should the Queen come back? Does infrastructural decay reveal that Uganda's national independence and sovereignty were a mistake after all?

Are Ugandans worthy citizens, or fundamentally unable to take care of the clean city and urban order left behind by the colonial power? Never mind that this was an order predicated on racial segregation; as Renato Rosaldo observes, "A mood of nostalgia makes racial domination appear innocent and pure."[10] In this formulation, a foreign gaze—a hypothetical Japanese ambassador—is deployed to offer a critical commentary on Ugandan citizenship, raising the specter of an inherent and essential racial difference—framed as the ever present trace of the village—that constructs an originary and radical incompatibility between Africans and urban life. This is not a new idea. On the contrary, colonial theories of essential racial difference were expressed through urban form, conceptually underpinning and materially constructing regimes of spatial segregation and racialized economic extraction during the colonial period around Africa.[11] Far from exceptional, these editorialists' views exemplify a recurrent trope in the Ugandan public sphere. In 2018, former ambassador William Naggaga suggested in his *Daily Monitor* column that a silent and growing majority of Ugandans excluded from the "fruits of independence" feel colonial nostalgia, even if only a few voice aloud "what others prefer to say to themselves out of fear of being laughed at."[12] These commentaries reveal the importance of understanding postcolonial cities as racially charged and racially constructed spaces. This is true not only because colonial cities were built and planned in relation to explicit racial hierarchies but also because race continues to influence the ways in which Kampalans inhabit, understand, and critique urban space and imagine urban futures.

Race is a silent, yet constitutive, element of the normative understanding of development that underpins and sustains dominant ideals of urbanization. In contemporary Kampala, racialization takes form around symbolic distinctions between urban-rural, city-slum, formal-informal, order-disorder, and discipline-chaos that reproduce white supremacy even as they do not mobilize an explicit white-black opposition. An unstated white-black opposition, sometimes finding voice in terms of a civil-savage hierarchy, is the master binary and underpins and animates the other oppositions through which the present is apprehended and the future imagined.

Of course, racial logics in Uganda have never been as simple as purely Black and white. As Mahmood Mamdani argues, the binary logic of colonial extractive and cash-crop economies was obscured by the vital role of Ugandan Asians who became the face of the colonial economy as brokers, compradors, traders, and managers, operating at the contact points between the Ugandan peasantry and global supply chains for both export cash crops and imported consumer goods.[13] Until the 1972 expulsion, Ugandan Asians occupied a precarious and contested legal position as permanently settled non-natives that troubled the racial management of urban space for both colonial administrators' civilizing missions and the first independent governments' pursuit of Africanization.[14] Invited back to Uganda by President Museveni as a source of foreign direct investment, Ugandan Asians have been critical to the neoliberal development of Uganda since the 1990s, along with new "foreign investors" from both South and East Asia, racialized minorities in the country welcomed for their cosmopolitanism, commercial networks, and entrepreneurial contributions to development.[15]

As in Ghana, as Pierre has shown, the postcolonial Ugandan state inherited the racialized structures of colonial rule constructed on the basis of white supremacy but mediated via indirect rule in such a way that the categories of ethnicity—deployed to rule the "native" population—emerged as a mode of incorporation that obscured hegemonic whiteness.[16] Colonial racial logics divided the Ugandan population into discrete tribes clustered into racial groupings opposing the "progressive" southern tribes engaged in cash crop production to the "martial races" of the north enrolled into the colonial army and police.[17] These categorizations were ideologically flexible. The "martial nature" of Acholi, Lango, and other "Nilotic" northern people, for example, was a convenient ethnological discovery that served as rationalization for recruiting these populations to combat anticolonial movements in the south.[18] The sophistication and cultural superiority of the southern tribes like Baganda, according to the Hamitic thesis espoused by early ethnologists and explorers like John Hanning Speke, was due to their racial origins outside Africa.[19] Far from deconstructing the racial logics of tribe and ethnicity, postcolonial governments have inscribed them as the foundation of national belonging through institutions like the Uganda Museum and Ndere Cultural Center that stage the nation as a mosaic of discrete ethnicities, each possessing a distinctive ethnic tradition embodied in its domestic architecture, dance, and dress. This form of ethno-racial governmentality is also reinforced under neoliberalism by nonstate actors, as ethnic traditions become fodder for marketing campaigns by a variety of commercial brands.[20]

Racialization in contemporary Uganda essentializes culture rather than biology as the locus of racial difference, emplotting these cultural differences onto the teleology of development. While the political sphere is reserved for a national elite, defined primarily through education, the majority of the population is relegated to the antipolitical sphere of development as subjects in need of uplift.[21] Michel Foucault's notion of state racism—the governmental elaboration of populations differentially included in the category of the human and the constitution of "others" as threats to be eradicated, managed, or transformed—offers a useful heuristic to understand the ways in which this mode of racialization operates by constructing cultural-cum-temporal divisions within the population.[22] As part of the post–World War II European intellectual project of theorizing the connections between European liberalism and Nazism, Foucault gestures to but does not properly account for the specific institutions of white supremacy in the colonial state or the slave trade, as many critics have pointed out.[23] Nonetheless, Foucault is useful for understanding how race and racism operate in postcolonial spaces insofar as he identifies race as a constitutive element of philosophies of history and modes of governance that separate out "the groups that exist within a population," essentializing and institutionalizing these separations.[24] In Nikhil Singh's formulation, state racism "has helped to create and re-create 'caesuras' in human populations at both national and global scales . . . that stigmatize and depreciate one form of humanity for the purposes of another's health, development, safety, profit, and pleasure."[25] Breaks within the population thus constitute certain groups as threats to the health and purity of others. More pliable than race theorized as biological difference, however, cultural racism offers the possibility of personal uplift and racial development. State racism thus informs not only the most violent modes of disposability but also the more ubiquitous life-making institutions of bio-power that attempt to disciplinarily manage threats to life by transforming one form of life into another—transforming villagers into urbanites, for instance. In countries ideologically structured around the notion of development, this means that racialization is not exclusively concerned with bodily difference but rather is located instead in behaviors, customs, attitudes, beliefs, objects, and lived environments through which progress is gauged.

As a narrative of teleological progress, development inherits the evolutionary framework of a movement from primitive to civilized, retaining this gap as the fundamental caesura in the population. Throughout the British Empire, development emerged as a practical and discursive means of resolving the contradictions of liberal imperialism, managing the gap between the universalism of liberal civil society and the necessary exclusions of colonial rule. It relegated racialized colonial subjects to an inferior, undeveloped, status, "providing, at once, justification for colonial rule and a normative template for colonial policies."[26] Colonial development implied "that the entire African population was slowly 'developing' into adult Europeans."[27] In settler colonial contexts, this notion of racial development via education and assimilation was articulated in both explicitly genocidal terms—"Kill the Indian, Save the Man"—and in the ranking of languages and cultures into tiers.[28] In India, Vinay Gidwani argues a cornerstone of this model of development was overcoming waste by "rationalizing" native forms of life, land tenure, and agriculture. While colonial civilizing improvement schemes—"the white man's burden"—have been widely critiqued as racist and racializing, the same civilizational projects recast as postcolonial development schemes have largely not been analyzed as racializing because racial difference has become less phenomenologically apparent as a structuring factor. Nonetheless, racialization resides as much in the nature of the burden as in the man carrying it.

The temporal construction of waste as an urban crisis caused by the underdeveloped cultural "village mentalities" of low-income Kampalans illustrates these racializing dynamics in which the structural problems of urban life are blamed on those most harmed by them. A KCCA officer participating in a 2014 corporate-sponsored community cleaning exercise in Nakawa, for example, identified the population's attitudes as the critical blockage in the city's waste management infrastructure: "Big-headed [residents] do not want to collect their garbage in one place to be collected by KCCA trucks." Echoing this sentiment, another participant quoted in the media coverage of the event diagnosed the problem as awareness: "People in slums litter garbage anyhow and do not mind. They constantly need to be taught about good hygiene and change their attitude." Material inequalities, like the inadequate provision of waste management infra-structures in low-income neighborhoods, are thus coded as reflections of essentialized cultural differences of mentality and personal development. Through sensitization campaigns like cleaning exercises, the culture of less civilized others becomes the primary terrain in which urban transfor-mation is sought.

During the 2011 cleaning exercise in Kasubi described in chapter 11, this process of biopolitical differentiation is manifested in the mayor's exhortation to the community to clean the market themselves for the sake of other residents who avoid it for its filth: "David lives at Nansana, but he cannot buy anything from Kasubi Market because of poor sanitation. . . . This is too bad for us. I felt so embarrassed." Here waste poses a threat to commerce, circulation, and the health of the population and, through middle-class fears of contagion, of the economy itself. In this hygienic logic, the connection between the professional and informal populations is rendered not as a liberal social contract but rather through the dangers of disease and the circulation of money, such that the urban poor appear as a threat to be contained through cleanliness, uplift, and responsibility or, alternately, through displacement and eviction.

In *The Wretched of the Earth*, written with uncanny prescience on the eve of Uganda's 1962 independence, Frantz Fanon anticipated that anticolonial nationalism could lead to "the revival of the commonest racial feeling."[29] Fanon warns that the national bourgeoisie will install white-supremacist racial hierarchy as the foundation of the new national development agenda: "Bourgeois ideology," he writes, "which is the proc-lamation of an essential equality between men, manages to appear logi-cal in its own eyes by inviting the sub-men to become human, and to take as their prototype Western humanity as incarnated in the Western bourgeoisie."[30] Rather than break with colonial racism, Fanon argues, anticolonial nationalism continues along its path of evolutionist (and extractive) developmentalism, both ideologically and economically. This white-supremacist ideological consensus facilitates neocolonial extrac-tive economies in which the new national economy cannot serve the needs of the people, instead remaining internally differentiated accord-ing to the needs it fulfills for the economy of the "mother countries."[31] The national bourgeoisie, Fanon warns, will simply take over the intermediary institutions of colonialism, leaving racism intact as a foundational ideology. Fanon points out the ways in which leaders of the newly independent African nations are liable to voice the racism of the colonists they replace: "The settler never stopped complaining that the native is slow. Today, in certain countries which have become independent, we hear the ruling classes taking up the same cry."[32] Fanon's analysis precisely anticipates a 2014 editorial by Beti Kamya, president of the opposition party, the Uganda Federal Alliance. Titled "Ugandans Are Poor Because They Are Lazy and They Celebrate Mediocrity," the piece argues that "the difference between Africans and Europeans lies in our values and attitude to work" and that poor voters who squander money on large families, communal events, alcohol, and witch doctors do "not deserve to be alive and leaders should say it to their faces."[33]

Despite having written a thesis on Fanon, violence, and the Mozambique guerrilla war for independence during his time at the Dar es Salaam University in Tanzania, President Museveni has taken up this colonial anti-Black cry with great verve in speeches given in the years since the Uganda@50 celebrations, berating Ugandans as lazy, dependent, and irresponsible and going so far as repenting on behalf of the nation for the "sins of laziness, indifference and irresponsibility."[34] Museveni's political thought, even as a radical freedom-fighting guerrilla, was structured by a cultural theory of backwardness and the necessity of development for the transition from savage to civilized society. In his autobiography, he described Uganda at the time he came to power in terms of lack, superstition, and backwardness: "Uganda was still very backward: millions were walking barefoot, underfed and suffering from worms, without adequate medical services; their mental world was still dominated by superstitions—to mention only a few of the woes that afflicted and still afflict the Ugandan and his fellow African. Economic and social structures lacked the ingredients to usher in the basic changes which could have transformed a primitive social order into a modern, civilized one."[35] Here the material conditions of the people—barefoot, malnourished, and diseased—and their culture—afflicted by a superstitious mentality—are evidence of a primitive underdevelopment requiring modernization and civilizing from the outside as they themselves lacked the vital ingredients.

This mode of emplotment, the temporality of development, is central to cultural racialization in contemporary Kampala. Class differences, such as those between the kinds of garbage disposal used in wealthy and poor neighborhoods (private collection companies as opposed to illicit dumping, for instance), are registered as differences in time, with poor people's practices described as residual. At a cleanup in Kawempe Division, Davis, a volunteer cleaner from a nearby NGO, told me that "these people [residents of a slum in Kawempe Division] still throw their rubbish carelessly." The word *still* is critical here. Here, as in Museveni's autobiography and in countless other uses I encountered in everyday speech, *still* relegates contemporary waste management practices to an earlier time occupied by the temporal-spatial other: "these people." People who discard rubbish "carelessly" have not yet progressed, their economic difference and infrastructural abandonment transformed into a difference in civilization. *Still* equally registers a spatial difference, referring to practices identified with rural villages, themselves only ever imaginable as exemplifying pastness. In this way, temporality is integral to the racialization of class difference. Here, class conflict is obscured by means of a temporal sleight of hand. The racialized urban poor are excluded from the present—where they make political claims to urban belonging through their assertions of presence and economic survival in the city—and relegated to the past.

This is not to say that the rural as such is totally devalued, either in national ideology or popular sentiment. Quite the opposite is true. Despite rapid urbanization, 75 percent of Uganda's population lived in rural areas as of 2019.[36] This rural base is framed as both a problem—people who need to be developed and converted to more productive forms of agriculture—but also as the bastion of national morality and images of the good life. This idea of essential African rurality informed the infra-structure of feeling of early-independence urban space in East Africa. Julius Nyerere's Pan-African socialist ideology of *ujamaa*, for instance, is premised on the assumption that "humanity [is] divided into discrete natural categories, each with its own traits and characteristics," as well as paths to development.[37] Within this ideology—which greatly informs Museveni's political thought—authentic African life stood in stark contrast to cities framed as morally suspect sites of consumption and luxury. The figure of the progressive peasant, a model farmer cum entrepreneur who has moved from

subsistence to commercial farming, rationalized his production, diversi-
fied his activities, and managed to resist the forces of kinship and speculation that threaten land fragmentation, remains at the core of official
policies for national development, like the thirty-year plan Vision 2040,
even if these visions are not supported by budgetary allocations. Museveni
seeks to cultivate this image of himself, returning frequently to his ranch
to tend to his cattle and branding himself "the man with the [herder's]
hat," as in a 2019 tweet of a photo of himself herding his beautiful long-
horn Ankole cows with a caption reading, "The old man with the hat has
this afternoon concluded supervision of the bi-annual exercise of deworm-
ing our cows."[38]

Many of the waste pickers and boda-boda drivers I met in Kampala told
me that they were saving to buy land in either their home areas or in the
rural districts surrounding Kampala where they could invest in poultry or
pig-rearing businesses. Others fantasized about buying a small truck that
they could use to transport food commodities like matooke from the rural
areas inhabited by their kin to the urban economies they'd come to know
as low-wage or informal workers. These types of entrepreneurial business ventures, profiting from linking city and the country, had much more
appeal to most of these young men than farming crops, labor dismissed
as mere "digging." The informal spaces of the city thus represent spaces of
opportunity in which many young men build themselves semi-agrarian
futures, trying to gather capital and connections in the city's cash economy
in order to gain a foothold in a rural economy that, far from an allochronic
idyll, is highly competitive and speculative. Rural land and life also function as relatively stable sites to save and invest, allowing precarious waste
workers strategies to counter the uncertainties and vicissitudes of urban
life. This has been a long-standing part of urban-rural entanglements in
Uganda. In his study of labor migration in 1950s Kampala, Walter Elkan
documents similar strategies and structures of desire.[39] Contradicting the
target-work thesis that suggested that African workers were only motivated by the desire to purchase specific consumer goods, Elkan found that
mid-twentieth-century migrant workers sought to save money in order to
invest in increasing agricultural productivity. Deprivation in the present,
in the form of inadequate nutrition and substandard living conditions in
slums, were part of this future-oriented savings strategy. Thus, the rural

itself is not abjected either popularly or by the state. Rather, it provides an explanatory framework—temporal disjuncture—for the practices and living conditions of the urban poor, who are abjected as matter-out-of-place disruptions of the geographic and social distinction between urban and rural that has, since the colonial period, functioned as the primary rupture *within* the African population of Uganda.

Uganda's fiftieth anniversary occasioned both national celebration and an existential crisis around the stalled and unmet promises of national development. Why is our infrastructure going to waste? Why is garbage piling up in our capital? Why are we "still backward?" For some public intellectuals, the dilapidated condition of urban infrastructure revealed the limitations of the postcolonial state and gave rise to sepia-tinged nostalgia for the colonial order of things. For Mr. Mulindwa and many other ordinary residents of the city, potholed, flooded, and congested roads, uncollected garbage, and sprawling slums were material manifestations of corruption, the scandalous theft of the wealth of the nation and international development funds by an unscrupulous political elite. For other political leaders and commentors, the chaos and disorder of the city stemmed from a lack of political courage to pursue developmental authoritarianism: the unwillingness of elected officials to take a hard line against informality and the selfish, lazy, and corrupt values of the urban population by collecting taxes and data, sticking to plans, and enforcing stricter laws.[40] In different ways, these perspectives on the merits of independence all share a commitment to development as the purpose of national sovereignty and a material achievement that indexes the quality of the state. In common parlance in Kampala, development is simply shorthand for the desire for a better life: for securing housing, healthy food, quality education, stable work, medical care, and meaningful social ties. While this constellation of popular desires gives the concept its affective power, in encounters with waste and cleaning projects the coloniality of developmental logics recasts material inequalities as moral failures and indexes of temporal differences. In the temporal politics of disposability, uneven development is culturalized and blamed on a population infantilized as uncivil children in need of education, awareness, and sensitization, or cast as backward villagers "still" unsuited to urban life.

14 Clean Hearts, Dirty Hands

At exactly 11 a.m. on November 11, 2011, Kampala City Yange, a new urban cleaning campaign, was launched with much fanfare, including a rendition of the Ugandan national anthem and a parade through the city with the lord mayor. A private foundation working (at the outset) in conjunction with the municipal government, Kampala City Yange (KCY) sought to foster a sense of volunteerism and to promote a sense of ownership and responsibility for the city by organizing cleaning exercises staged on the last Saturday of each month in a different neighborhood in the city. KCY appealed to Kampalans' patriotism, calling on them to join because "we are convinced that the state of a capital city of any nation plays a big role in shaping the image of the entire nation and it is therefore imperative that we as city dwellers all play our part in building a greater city than the current one."[1] Choosing 11 a.m. on 11/11/11 to launch themselves into the public, they aimed to inaugurate a new era of unity and oneness between the municipal government and the urban citizenry around the shared goals of beautification and transforming Kampala into a greater city.

While it sought to inaugurate a new time of civic participation and cleanliness, KCY harkened back to Ganda traditions to do so, seeking to revive a spirit of patriotic nationalism and to reignite a "lost culture of volunteerism."[2] The primary referent for this lost sense of community was *bulungi bwansi* (for the good of the land/country), a Ganda community event. This was a three- or four-hundred-year-old practice, Godfrey Katete, one of KCY's founders recounted, whereby everyone in the community would come together at the sound of a particular drum beating the call "sagala agalamidde" (literally "I don't want anyone lying down," meaning "come and work with us"), summoning people to clean up, build roads, and work for the betterment of the kingdom. "City Yange is a modern version of that," Godfrey explained. But, he joked, "This is a city and walls are up around people's houses, you'd need VERY big drums!" The new drum, he showed me, is a music video by local singer OS Suna, commissioned specially for KCY.[3] The slickly produced Afropolitan video restaged the bulungi bwansi tradition, drawing its beat from drummers in front of the Kabaka's residence, calling out "sagala agalamidde" with footage cutting between scenes of cleaning exercises, ordinary urban life, and shots of OS Suna sporting a traditional bark cloth robe, singing, dancing, and cruising around Kampala in a BMW convertible over the chorus "*longosa city yange* (cleaning my city), keep it clean and green." Blending past, present, and future, the video expresses KCY's aspiration to foster a popular culture of cleanliness for Kampala.

The ancient practice of bulungi bwansi had died out by the 1990s, Godfrey explained, because of the decades of national trauma, the growing urban disorder, and the "mentality of laissez-faire." Today, Ugandans have lost this spirit, and "if you say anything about volunteerism, people would think of a white person like you coming to Uganda to save the orphans." I encountered this correlation between whiteness and volunteerism consistently during my time in Kampala. "Where are you volunteering" was often among the first questions Ugandans would ask me when we met. As Godfrey commented, volunteering was coded as a foreign activity, something for young white people to do. KCY was turning this foreign and racially "other" practice into something young, trendy, and aspirational by inviting popular musicians and sportsmen to participate, by communicating through music videos, and by inviting student associations to join their cleaning exercises. KCY promoted cleaning the city as a trendy weekend activity, part of a novel lifestyle with cool music, hip clothing, and local celebrities. This linkage between cleaning and trendy lifestyles discloses the importance of cosmopolitan sensibility and desires within the contemporary infrastructure of feeling of developmental respectability through which urban transformation is being promoted and inhabited. By making cleaning cool, KCY sought to overcome the perception of volunteering as a foreign, white activity and mobilize the city's upwardly aspirational youth for urban transformation. But in doing so, these cleaning events interpellated cleaners into a humanitarian structure of feeling predicated on a racializing gap between savior and saved.

Responding to the viral phenomenon Kony 2012, Nigerian novelist and critic Teju Cole sardonically joked that "the fastest growth industry in the United States is the White Savior Industrial Complex," commenting that "the White Savior Industrial Complex is not about justice. It is about having a big emotional experience that validates privilege."[4] Cole's scathing view of sentimentalized transnational humanitarianism points to the dangers of "good hearts" that see and feel suffering but cannot think of its causes. His white savior emerges in contradistinction to a needy, hungry, benighted African whose life becomes little more than a mise-en-scène for heroism. This split between "lives to be saved and lives to be risked"[5] takes less dramatic but similarly racialized transnational forms in mission trips, fundraising campaigns, and voluntourism that construct affectively charged encounters between the global poor and well-meaning volunteers from the global north.[6] Privileging the production of positive affect over any structural analysis of the causes and consequences of poverty, these humanitarian-developmental regimes locate and produce knowledge about the problem of poverty within discrete bounded territories like slums and camps and within the attitudes and bodies of the poor themselves, constructing these geographies as consumable experiences through which white volunteers can forge novel identities and bolster their professional credentials.[7] The transformative power of white saviorism was satirized by a 2014 heading in *The Onion*: "6-Day Visit to Rural African Village Completely Changes Woman's Facebook Profile Picture."[8]

While the political dynamics of these encounters between volunteers and the global poor are readily apparent and easy to satirize, the ways in which they establish a racialized set of affective norms and expectations that inform the humanitarian governance of poverty and cleanliness within emergent national class formations is less visible, though no less central. Humanitarian sentiment, and the gap it posits between the savers and the saved, the redeemer and the redeemed, are a central component of the structure of feeling of Kampala's cleaning exercises, mediating the encounters between upwardly mobile volunteers and the marginalized urban populations to whom they "give back." How are these ruptures culturally constructed and materially enacted? How do well-intentioned volunteers become caught up in the reproduction of racialized class hierarchies despite their aspirations to uplift and give back? How do humanitarian affective investments, and the regime of care that they constitute, contribute to disposability?

"Kampala has the potential," the KCY organizers wrote, "to be the most beautiful garden city of Africa. Kampala could become the most memorably exciting place to live and work in. Kampala could become the most happening city experience for many people."[9] The KCY campaign was designed to feel good, to build on the city's potential and inspire residents to come together, to participate in building and valuing public space, and to provide opportunities to materialize hope by "confronting littering, waste management issues, cleanliness, environmental concerns and beautification."[10] After we met and talked at a few cleaning events, the campaign organizers, Ezra Kalungi and Godfrey Katete, invited me to their well-appointed offices to talk about the ideas behind and the future ahead of KCY. Sitting in their conference room, with a window opening onto a spectacular view of the city's hills, Godfrey and Ezra talked me through a well-rehearsed PowerPoint presentation explaining the KCY campaign. KCY focused on cleaning, Godfrey explained, because when they asked around, "the most common issue that was raised was, 'There is too much garbage all over the place; the city is filthy!'" Cleaning would have a range of important "trickle-down effects," as Godfrey put it, reducing malaria by eliminating stagnant water, fighting crime by "removing dump sites where criminals hide to prey on the community," cleaning the air to allow children to play freely, and building a sense of pride and peace within communities by getting everyone out to work together and meet each other, "just as they would at church."

Godfrey and Ezra used their connections across the city's evangelical churches to get the message out and reach potential volunteers. George

Mugarra, a frequent volunteer I first met at a KCY clean up in Kabalagala in October 2012, heard about KCY in this way. George earned enough at his marketing job with the telecom giant MTN to identify himself as part of the middle class but was looking for new work so he could move out of his aunt's house. Despite his relative comfort, he identified as part of a generation of educated and aspirational Kampalan youth frustrated by the lack of meaningful work or economic opportunity. He channeled his energies into church, regularly attending the prestigious Watoto Church, an evangelical church famous for its positive message, its popular radio station, and its celebrity-packed congregation, including prominent public figures like the KCCA executive director Jennifer Musisi. He first heard of KCY at Watoto, when Ezra, another prominent member of the church, announced the campaign. "I was excited to see someone come up with City Yange. . . . It's one thing to have, to have it in mind, but another thing to get your hands dirty. Most people want a clean place, but they don't want to get dirty. But I decided to get into action." His attitudes toward the event and the overall project of cleaning up Kampala resonated with nearly all of the other young people I talked with at City Yange events and in follow-up interviews. George came to Kabalagala eager to dive in, explaining to me later that "I didn't mind [getting dirty] because I purposed to do that. In fact, it felt good, being part of the solution, not part of the problem; that's a great feeling!"

George's great feeling resonated with the overall affect of KCY events in general and the Kabalagala exercise in particular. A few minutes after we first met—in the car park of Kampala International University, the staging area for the month's cleaning—we listened to a speech launching the day's activities. A leader from one of the partner organizations (a youth group all wearing T-shirts with the slogan "I am Activista") thanked us all for being there "to give back to the community" and warned us, "We are going for those smelly things. Don't fear! They smell, but we shall do it, go and pick! Let us go to the smell." Ending with a call "city *y'ani*?" ("whose city?"), to which we shouted back "city yange!" ("my city!"), reaffirming a personal possessive relationship to urban space. The next speaker, from the Uganda Red Cross, told us, "We are here to show the community that people outside care for them, care about their health. Let them join us, but let them join from the heart, in the same way you have volunteered." Here, being part of the solution and volunteering from the heart remade dirt as an index of care, responsibility, and future making.

The cleaning event's infrastructure of feeling guided moral attention to Kabalagala's filth, where volunteers would encounter abject urban matter and transform it into feelings of pride, pleasure, and care. Listing obviously repulsive things in garbage was a common rhetorical move when organizers and participants, as well as journalists covering cleaning exercises, sought to emphasize the disgustingness of waste and the otherness of spaces in which it is found. The most disgusting things people emphasized typically had connections to bodily fluids; specific items included feminine hygiene products, diapers, condoms, and "flying toilets" (plastic bags used domestically as nighttime toilets that are then thrown out onto streets and roofs). Exaggerated repulsion from such radically "other" substances is, as Stallybrass and White argue, an affective response through which respectable subjects are formed within sanitary modernity.[11] Being part of the solution required activating, but then suspending, volunteers' typical forms of avoidance and embracing filth, seeking out, rather than dodging, "those smelly things." In transgressing the taboo on filth by getting their hands dirty, volunteers ultimately reaffirmed the distance between them and the places and people being cleaned. In contrast to the filth of the problem—framed as a product of uneducated, impoverished, and not-yet-empowered slum communities—and in marking the urban poor as morally culpable for their abject living conditions, the filth of the solution is understood as an effect of the labor of purification. This filth is generative rather than destructive, producing the feeling of citizenship. This logic is stated explicitly in a 2012 newspaper article on KCY: "There is no flinching; the dirtier the garbage the more honorable it is to clean it up."[12]

Good citizenship is thus an effect of particular kinds of work (dirty, degrading) done for particular reasons (duty and love rather than desperation) and with a specific relation to money (free from the impure motives of remuneration). These distinctions separate volunteers' work from the kinds of cleaning done on a regular basis by slum dwellers and the urban poor. The assumption is that "typically in Uganda, recycling is done out of desperation, run by poor women operating in slums."[13] Done out of desperation rather than desire, this kind of dirty work does not constitute citizenship but serves only as evidence of degradation, not environmentalism or respectability. Unable to rise above mere survival, the woman in the slum cannot be considered a real citizen and so remains merely a subject of development, the constitutive other of the volunteer-citizen.

Dealing with garbage becomes honorable when it entails overcoming an embodied attitude of aversion, the flinch. The Christian influence on this structure of feeling is apparent, even if it was not explicitly invoked within the surprisingly secular discourse of KCY. As in Jesus Christ's practice of foot washing, cleaning garbage positions the volunteer as "one who serves."[14] Engaging in a task that is simultaneously menial and intimate reminds volunteers of the value of humility. The purity of volunteers' spirits and intention suffices to overcome the filth of the matter they encounter: their ability to handle filth serves as testament to their altruistic dedication to community service, a secular iteration of Christian spiritual cleanliness.[15] In contrast to volunteers' assumed aversions, and the moral logic that purifies their encounters with filth, part of the diagnosis of the city's garbage problem—a diagnosis shared by elite and impoverished reformers alike—is that too many Ugandans do not feel this repulsion. Far from it, they are far too comfortable working alongside dirt and debris: "We want to make the population develop a phobia for garbage and hence remove it in the most appropriate way every day," a councillor for a flood-prone low-income area in northern Kampala told a reporter at a May 2012 cleanup.[16] With proper attitudes, the population would cease disposing of their rubbish "in the most bizarre of ways—including dumping it in drainage channels, stuffing it inside *buveera* [plastic bags] before dumping it on the roads."[17]

Along with documenting these "bizarre" practices, press reports on cleanups highlight the abject living conditions of Kampala's slums. As one journalist, Jonathan Adengo, wrote:

> I was shocked to find out how business thrived alongside the rubbish. On one side, stood a lady selling pancakes and chapattis and baking more, just next to a pile of rubbish. And just across the flooded muddy road were shops and a small pub very busy attending to their customers near blocked drainage channels and rubbish dumped by the side. As we went deeper inside, we uncovered another sorry state marred by blocked drainage channels and poor waste disposal. Walking through these communities would bring you to a conclusion that life is cruel indeed. The hygiene of the place was sickening and to imagine that people lived there! . . . The flooded roads and the

sight of people carrying on their business activities next to a pile of rubbish was a worrying sight. Some people had even turned the incomplete houses to sites for rubbish.[18]

This illustrative passage repeats many of the central tropes of middle-class reformist reportage on the living conditions of the urban poor in the early industrial era.[19] It focuses on the promiscuous, indiscriminate mixing of activities, the abominable proximity of people to their waste, the lack of functional separation of space, the mingling of filth and food, and the broader immorality associated with these spaces. What is especially abhorrent to Adengo in this passage is that the residents do not even seem to mind or care about their conditions, happy to continue their economic activities amid the filth. Adengo's response is a sick feeling and shock, not only at the conditions but at the very idea of this form of life. His response is evidence of his cultivated, civilized being and stands in contrast to the degradation and depravity of the unimaginable others inhabiting the slum. The sentimental logic of developmental respectability, however, renders these conditions as essential and natural effects of the culture and ignorance of the urban poor, to be overcome through sensitization. The infrastructure of feeling forestalls empathy and solidarity. It discourages a political economic or historical analysis of the living conditions in Kampala's low-income settlements, turning instead to a moralistic cultural-cum-racial othering of economic inequality. The slum's conditions, however, provided the journalist with an "eye-opener" and an opportunity to "give back to the nation."[20] He writes, "It was an honor for me to be part of a team I believe was cultivating disciplines of hygiene and love for our city as we celebrate 50 years."[21] Moreover, he notes, "being a part of this exercise made me feel proud as a citizen."[22] Describing the cleaning itself, Adengo describes how the team "moved inside the communities armed with cleaning gear, spades, easy pick hand tools, in [a] true spirit of love for our city and country."[23]

Cleaning takes on a masculinist and militarized tone, with "armed" cleaners penetrating "deeper inside" the (in turn feminized) community. Another reporter, covering a cleanup in August 2012 similarly frames cleaning as an act of aggression: "It was [all] business as the cleaners *attacked* the channels, scooping out all sorts of garbage recklessly dumped

Figure 7. Residents and volunteers clear a drain during a cleaning exercise in Katanga. Photo by Jacob Doherty, 2014.

by area residents. Thank God for the Kampala Yange mighty campaign!"[24] These militarized metaphors frame urban transformation and the struggle for developmental respectability as a battle against chaos and disorder, a battle that is, to some extent, being fought against the nature of Ugandans themselves. Despite the presence of military and police vehicles at every major roundabout and green space in the city, the forces of law and order were perceived to be on the defensive as "vendors stalk every street, [and] hawkers hold every block," in the words of social commentator Andrew Mwenda.[25]

Godfrey recognized that cleaning the city would be disruptive, telling me that "the thing with all this filth that was filling the city [was that] people had lived with it for so long that it had become normal. Disorder had become the new order, and anyone trying to change that was being disorderly; that is why people are resisting Jennifer [Musisi]. This woman, who does she think she is? We've grown up here; this is our home!" Here, Godfrey acknowledges that many Kampalans have been able to sustain their lives in the city because of, not in spite of, its disorderliness. Nonetheless, the urban transformation they seek to support begins with the reassertion of order: "The transformation . . . is just being able to bring order back in place. Once the order will be there then it will be a lot easier to implement all these other things." Here again, City Yange was simultaneously backward and forward looking, seeking to bring order *back* in order to secure the ground for a suite of other reforms. KCY cleaning exercises took place at the point where communities' habituated disorderliness meets volunteers' desires for orderly futures. Godfrey recounted an early cleaning event in a slum named Katogo, where he identified the power of cleaning to transform the mindset of the urban poor:

> By around ten o'clock people are hungry, so we figure they're looking for something to eat, and there are these women selling bananas. People are buying these bananas because they are hungry; they start eating them. And in this culture when you're done eating, you just throw it away; you don't even think about disposing of it. Now, our concern is to change their mindset, to say: "You can't just litter anyhow; that isn't an environment you want to live in. You want to live in a nice clean environment." You can't just go to their door and tell them "you don't do that." They've lived that way, and disorder has become their order. You [i.e., Godfrey] look disorderly; you are

the one who is bringing disorder! So in the middle of this cleanup they start throwing peels, but there is a self-realization among a couple of them: "Wait, wait, why are you throwing those things? We're here to clean up! So it defeats the purpose of us coming here; we should've stayed in our houses if we're going to throw the peels anyhow." So they realized, "OK, you're right, you're right about this." They started picking up these peels and put them in the bags that they'd brought. Right there the behavior change is happening in real time!

For Godfrey, behavior change is the key to securing meaningful urban transformation, but this must be accomplished through immersive pedagogical encounters like cleaning exercises where the labor and materiality of urban development are made evident. Urban transformation starts with internal self-transformation sparked by events that make residents realize the links between their individual behavior and their environmental conditions. For Godfrey, simply telling people what to do would be the disorderly thing to do, as it would disrupt communities' existing patterns. However, staging events ideally allows for self-realization, manifesting the disjuncture between the work of cleaning and the "culture" of careless disposal. This is a transformation that begins with individual moments of personal reflection and dawning awareness in a few members of an otherwise homogenous "community" who, in turn in this apocryphal narrative, are meant to themselves take on the mantle of developmental respectability as role models and leaders who transform the rest of their community.

Despite the role of elected officials as honorary cleaners at the forefront of most events, politics was generally viewed as anathema to this project of personal responsibility, community togetherness, and civic duty.[26] Given the autocratic context of Kampala and the widespread disenchantment with government officials and political representatives as corrupt "money eaters" not concerned with the public good, it makes sense that KCY and the volunteers who flocked to it wanted to carve out a space where they could do moral work outside of politics. The very word *politics*, in Kampala, was generally limited to refer to the venal realm of electoral representation, party jostling, and behind-the-scenes scheming. As such, politics was readily dismissed as selfish, inconsequential, and artificial. Doing development through projects like city cleaning exercises was moral precisely because it was outside of the political sphere, even as politicians were invited to reap the moral harvest that cleaning could bestow. Volunteers I spoke with were less interested in the presence of politicians than in the opportunity to give back and to enact social responsibility. In this sense, developmental anti-politics was a refusal of petty electoral scheming for positions of patronage through which individuals could eat at the expense of the public. Venal politicians represented an important part of the figurative infrastructure through which the moral position of the volunteer was constructed—politicians' corruption cast against cleaners' purity. Despite the rhetoric of unity and oneness saturating the events, contrasted with the divisiveness of politics, this form of developmental respectability—constructed as altruistic community mindedness—introduced a rupture into the population, constituting volunteers as agentive citizens working to improve and uplift passive communities.

This racializing rupture is reproduced more broadly in the class distinction separating the sphere of politics from that of development. Respectability shapes this distinction insofar as proper and improper speech are policed as ways of controlling access to the public sphere. As Florence Brisset-Foucault shows in her analysis of radio debates on Kampala's airwaves, these norms constitute a division within the nation between an elite realm of politics on one hand and a popular realm of development on the other.[27] Respectability and the norms of proper conduct constitute this form of class distinction while simultaneously offering individualized means of upward mobility that obscure and naturalize bifurcation. The disavowal of politics is paired with a highly individualized ethic of social responsibility that takes shape in sporadic, eventful, and affectively charged spaces like that produced by KCY. The amorphous desire to give back to an anonymous collective, figured as the community, replaces the Ganda moral economy of personal reciprocity and material exchange between hierarchically configured patrons and clients that both reproduces and mediates social inequality and affirms the personhood of recipients within a moral social order.[28] In addition to disavowing politics, cleaning campaigns also disavow this mode of interdependent solidarity and replace it with individualized gestures of care and individualizing respectability politics of moral uplift and personal responsibility.

Cleaning events clearly stage class difference. But a purely class-based analysis of these encounters obscures the ways in which this difference is informed by racialized assumptions about embodied aversions, normalized living conditions, and attitudes to filth. Differences in wealth are less significant than those in behavior, and a strictly economic understanding of the politics of cleanliness is complicated by the ways in which wealthy Ugandans are equally implicated in critiques of the city's garbage problem. There is a common classist assumption that the rich ought to know better. For instance, an MP, describing her hopes for an anti-littering bill she coauthored, explained that it would "help reverse a tradition of indiscriminate disposal of garbage where you find even those in posh cars who you expect have a possible higher standard of civilized conduct, casually tossing used beer or soda cans, or banana peels through the window as they speed along!"[29] Within this imaginary, car littering in particular takes on a highly charged form, the assumption being that people with enough money to own a car should have the proper dispositions toward disposal. Moreover, the diagnosis of attitudes as the obstacle to cleanliness implicates the population at large: "The majority of people in the community, both the elite and the illiterate [here class is rendered as a difference in education], are not concerned about where the garbage ends, provided it is out of their immediate environment. That is why garbage is dumped in drainages and on roadsides."[30]

While there is the assumption that elites should be more responsible for maintaining the cleanliness of the city, they are failing to do so because of "backward attitudes," selfishness, and laziness. Likewise, wealthy Ugandans' disregard for proper building and planning procedures means that even wealthy neighborhoods earn them opprobrium as "rich people's slums."[31] In this formulation, disorderliness and improper disposal do not map neatly onto class divisions. Rather, they are explained in cultural terms, as residual attitudes that belie the continued presence of backward African village life across the class spectrum and the inability for even wealth to truly overcome the essential cultural difference between Africans and respectable forms of life. What is at stake in cleanliness and anti-littering campaigns is thus the distribution of civilized behavior. Civilized behavior attests to the truth of the subject "running deeper" than the circulation and accumulation of money.

The "great feeling" at City Yange cleanups was a product of a distinction between the problem and the solution that takes human form in the figures of the community and the volunteer. Communities *are* dirty, whereas volunteers *get* dirty. KCY events simultaneously constructed (and heightened, as shown below) this distinction and sought to overcome it. The idea behind the cleanups was that community members should be inspired to join in. George's only criticism of the event, however, was that "the people where we went to clean were not as involved as we were, which I think is not good." This criticism was repeated by every volunteer I spoke to at every cleaning event, no matter the actual level of community participation. At the cleaning exercise in Kabalagala, the distinction between activists and the community was especially pronounced because of the free T-shirts that were distributed to volunteers in the parking lot staging area. While the speeches were going on, a KCY staffer passed out striking black T-shirts with the City Yange logo and the image of a man (stenciled in bright green) wearing what resembles a biohazard suit as he sweeps up in front of a row of skyscrapers on the front, and the slogan "THIS IS MY CITYING ROOM, DON'T LITTER HERE" on the back. When we walked through Kabalagala, all the volunteers (and even one or two of the police escorts) were dressed in the same black T-shirts and bright white rubber surgical gloves distributed by the Uganda Red Cross. This created the distinct impression that cleaning was a marked activity and that the exercise was limited to the properly attired volunteers. Assuming that I was among the event's organizers, a Kabalagala resident making omelets and chapattis at the side of the road asked me for a T-shirt so he could join in. When I explained that I didn't have any, he asked why he should join in if he wouldn't get the same benefit as the volunteers.

While cleaning exercises are a chance to "give back" to an unspecified community, this entity remains elusive. Invoked in the abstract, community is the constitutive other of cleaners' subjectivities as volunteers, meant to materialize in ways that affirm their generosity. But residents and workers in the neighborhoods being cleaned by KCY were often reluctant to participate in the humanitarian ritual being staged in their spaces. In March 2013, at a cleaning exercise in Katanga this dynamic recurred. Katanga is a small but densely occupied neighborhood crowded into a valley bottom on contested land between Makerere University

and Mulago Hospital on the neighboring hillsides. Although they partnered with the KCCA—who sent trucks as well as a crew of sanitation
workers and enforcement officers—this turned out to be one of the final
KCY cleaning exercises. Having lost their major sponsor, KCY could not
afford to print and distribute T-shirts. They managed to secure a dona-
tion of tools (brooms, digging forks, and shovels) and gum boots from a
bank (whose parking lot was used as a staging area). Without T-shirts,
the distinction between the volunteers (in this case members of the local
Rotary Club) and the community was less visibly pronounced. This was
the most well-received cleaning event I attended in Kampala. Hundreds
of children from the area joined the cleaning, running through the narrow
muddy passageways gathering litter into plastic bags. A local leader was
grabbing passersby by the arm and humorously hectoring them to join in,
pointing at me in particular to kiddingly shame them with the fact that
"even this *muzungu* [white person/foreigner] can get involved. *City zaffe*
(It's your city)!" Nonetheless, Alex, a young Rotarian I met at the event,
complained that the community was not involved. Along with a friend,
he took it upon himself to lead the cleaning of the wider—though still
impassable for garbage trucks—commercial thoroughfares leading from
the main road down to the drainage channel that runs along and defines
the center of Katanga.

Alex went door to door, demanding that the person working in each
small establishment (laundries; two-pot restaurants; phone repair stalls;
street-food stands; and retail kiosks selling soda, snacks, and airtime) stop
their work and sweep up the immediate area surrounding their premises.
These workers resisted, explaining that they were not proprietors and were
not getting paid to clean up. One owner told Alex that he paid taxes to the
KCCA to clean, so why should he have to help? Despite these complaints,
Alex managed to cajole most of these informal shopkeepers to do at least a
cursory sweep, gathering the accumulated debris (broken flip-flops, empty
milk packets, plastic bags, airtime scratch cards, broken glass bottles) from
underneath and around their stalls. At the top of the thoroughfare we
found a small wooden evangelical church built up on stilts with largely open
sides. Characteristic of churches in the poorest neighborhoods of Kampala,
this church was a work in progress, the construction process serving as
material evidence of the church's spiritual growth. Alex interrupted the two

men inside speaking loudly in tongues and demanded that they clean up all the garbage that was collected underneath the church floor. The two men filled up Alex's wheelbarrow, and he and I pushed it back down the hill to empty it out onto an enormous heap of garbage that was growing at the center of a dirt soccer field next to the central drain where a garbage truck was expected to come and collect it. As we emptied the wheelbarrow, I asked Alex what motivated him to take part in the cleaning exercise. He explained that as a Rotarian he is "always doing corporate social responsibility" because he "wants to improve life in these communities."

Talking with other volunteers I found out that most did not come from Katanga itself but were members of various organizations based in the slightly more prosperous areas on the surrounding hills. They had received funds from the municipal government for their own community development projects and were expected, in turn, to attend these events. I followed a group of young men I'd been chatting with as they headed off to look for filth. We stopped at a clogged drain, assessed the situation, and decided we would begin here. Soon, I found myself straddling an open drain and wielding a hoe, joining the four young men in the task of desilting and clearing the drain. It quickly became clear that urban planners' distinction between sanitation and solid waste is a precarious accomplishment and a privilege not enjoyed by the majority of Kampalans. Standing awkwardly over the drain we shoveled out garbage, mud, broken flip-flops, sewage, plastic bottles, and plastic bags. Dozens and dozens of plastic bags. The young men saw this material as evidence of the moral failings of the Katanga community. "How can they just throw this stuff here?" one of them asked rhetorically.

This idea structures much of the policy approach to the garbage problem in neighborhoods like Katanga. The idea is that the urban poor are ignorant of good disposal practices and need to be educated—*sensitized* in the common developmental parlance—to behave properly. But this ignores the importance of the city's topography. Kampala receives torrential downpours during the two rainy seasons, and residents of more affluent uphill neighborhoods take advantage of the rain to dispose of their own waste. It's common to see sacks of rubbish by rainwater-draining channels across the city, left out for the rain to take away. But there is no such place as Away, and since rubbish runs downhill, it accumulates in low-lying areas like Katanga where it clogs drains, causes flooding, stagnates water where malarial mosquitoes breed, and leads to sporadic outbreaks of cholera and typhoid.

We kept shoveling, accumulating a few sacks full of soggy, messy trash and silt that other volunteers loaded into plastic sacks and carried off. Tired of hunching over and eager to jot down some notes, I took off my gloves and followed these volunteers to see more of the cleaning exercise. The municipality had installed a dumpster temporarily in the area, and the volunteers took the sacks of trash there to deposit. Residents were taking advantage of the cleaning exercise and the temporary infrastructural improvement to empty their own bins, coming out to throw out their rubbish. A municipal worker used a shovel to pack the rubbish into the dumpster to make room for more.

I was taking photos and writing notes when I heard a middle-aged female volunteer ask, of no one in particular, "Where is the community?" "What do you mean?" I asked, and introduced myself. Her name was Grace; she was here with a community health team that does care monitoring for HIV/AIDS patients in a neighborhood a thirty-minute walk away. "They are not here—why are these people hiding? They stay indoors and refuse to come and clean, they leave it to us to clean for them," she explained. "They are lazy," added a passerby from the same organization. "No, they fear that they will get fines for being unsanitary," offered another. "Maybe they are working," I suggested, knowing that the city's informal economy—the source of income for the majority of Kampala residents—does not stop for Saturdays. "No, these people cannot work, look where they live, they are just lazy," Grace responded, settling on the passerby's explanation. Grace then led me to another area where more volunteers were standing, urging me to "come and see" the effects of residents' laziness. A municipal worker had discovered an illicit dumpsite in the alleyway between two rows of rentals and was clearing it out with a shovel. Here again, developmental respectability forestalls a political-economic understanding of labor and the environment, invoking moral failings and essential personal inadequacies to explain economic and infrastructural inequalities. Grace and other volunteers expressed pity, shame, and disappointment, but because it reaffirms a discourse of personal responsibility for environmental conditions, the infrastructure of feeling forecloses other affective responses such as indignation at the conditions of munici-pal abandonment.

KCY was described as a "lifestyle initiative" that sought "to promote a sense of ownership of the city by the city community." Ownership is cen-tral to KCY's philosophy. Their name makes direct reference to posses-sion: *city yange* meaning "my city." Ezra explained the double implication of *yange* as both endearing and aggressive: *Yange* "can mean 'mine' in a very endearing way, you know a mother to a child. But it can also mean . . . a kind of possessiveness, 'this is my city, you're not going to mess with it!'" *Ownership*, in this use, is primarily affective, referring to a sense or feeling of ownership rather than a redistribution of property. A slide in the PowerPoint stated, "All these experiences that you derive out of a city setting are 'a share of ownership' on your part depending on whether you are a resident, a worker, beggar, student, thief, shopper, visitor, or tourist to Kampala." In this formulation, ownership is a *feeling* that derives from experience and thus becomes available to a surprisingly inclusive cast of characters, figured as shareholders.

Lockean property theory is based on the idea that property is the result of mixing labor with the material world, improving the condition of things to lead to private ownership of them.[32] The moral logic here is that owner-ship confers responsibility and affective attachment that sustains care for property. In KCY's articulation, the relationship of ownership, improve-ment, and feeling is inverted. The campaign encouraged "every stake-holder to *own a responsibility*."[33] A feeling of ownership as responsibility (or even responsibility itself as a feeling one can own) was cultivated as a means of extracting labor. This occurs by mobilizing developmental respectability and its sentiments: a sense of responsibility means taking pride in a place and working hard to avoid the feelings of shame or embar-rassment that elites project onto it. This sense of affective ownership dis-avows the lived reality of property relations in Kampala's slums where the majority of residents are tenants with little to no security of tenure and, thus, according to the traditional Lockean logic of bourgeois property, no incentives to care for their environments.

Cleaning exercises sought to produce propriety through the racial logic of respectability, but without the concomitant relations of property that traditionally secure proper, respectable ways of life. Equally, KCY's developmental respectability placed the onus on volunteers and communities to improve cleanliness without extending responsibility to the actual property owners collecting rents from substandard rental units lacking adequate sanitary and waste disposal facilities. While the residents of neighborhoods I interviewed in the weeks after cleaning events often expressed their frustration at the futility of caring for shared spaces like streets, communal latrines, and drainage channels, they did reveal intensive regimes of domestic cleanliness over those spaces they did control, even as tenants. Many tenants I interviewed feared the entanglements of cleanliness and displacement. A resident of Kabalagala told me, the week after a cleaning exercise, that "if it gets clean here, then maybe they will want more [money] to live here and we will have to shift." I asked where she would have to move. "Further away [from town and the perceived center of the city's economy], even to Ggaba side," she replied, referring to a lakeside neighborhood as far from the center as it is possible to be. She was not optimistic that the conditions there would be any better than in Kabalagala: "We poor people cannot stay where it is clean." Far from realizing ownership and the implied stability and responsibility it entails, cleaning events presaged displacement, giving rise to feelings of vulnerability, temporariness, and exclusion.[34] The constitutive other to developmental respectability's structure of feeling, this resigned sense of inevitable dislocation was a common affective response to disposability. While for volunteers cleaning events felt good, fostering a sense of civic pride and personal responsibility by uplifting and giving back to communities, for recipients of this eco-humanitarian gift, they exemplify the potential violence of care.[35]

Disposability is produced not solely through unknowing, exclusion, and abandonment but also through affective investment and moral modes of inclusion that individualize responsibility and naturalize the uneven burdens of waste worlds.

The humanitarian structures of feeling that these acts of public cleaning are embedded in and reproduce are not inevitable or universal. Rather, they stand in stark contrast to other modes of cleaning and polluting around Africa that have politicized waste to mobilize feelings of solidarity, insurrection, or spiritual renewal. In Dakar, for instance, the Set/Setal movement of the late 1980s combined public cleaning with art, music, and street protests to critique corruption, infrastructural abandonment, and perceived moral decay and to animate and give purpose to a generation of youth stranded by structural adjustment and gerontocracy.[36] In Cairo, amid the protests and occupations of Tahrir Square in 2011, activists used urban cleaning to mark the resignation of Mubarak. Celebrating their victory, "Egyptians wanted to clean—to cleanse—the entire country, to rid it of trash, of the old regime"—and did so with brooms in hand to construct a sense of a shared moment, a collective investment in public space, and revolutionary agency.[37] In Cape Town, protestors have harnessed the disruptive abject powers of filth, spectacularly spreading shit at airports, highways, galleries, university campuses, and other centers of political, economic, and cultural power. Interrupting the smooth, sanitized circulation of the city, activists brought attention to ongoing racial inequalities in service provision, exclusion from "the grid of modern life," and the sensory conditions of everyday life of the urban poor.[38] Rather than Teju Cole's "white savior industrial complex," in which the production of positive affect for a racialized savior—constructed in contradistinction to an abject saved—obscured any consideration of the structural causes of poverty and inequality and the material realities that reproduce them spatially, such insurgent cleaning mobilizes waste, dirt, and the infrastructures that manage them to make visible, problematize, and disrupt the structural violence of sanitary modernity and disposability.[39]

Conclusion

SURPLUS, EMBODIMENT, DISPLACEMENT,
AND CONTESTATION

The Uganda Railway transects Kampala, running east to west through the heart of the city. The rails bear witness to the multiple layers of infrastructure sedimented here over time. Conceived in the late nineteenth century during the scramble for Africa as a means to consolidate British territorial claims in East Africa, by 1901 the railway connected the Kenyan port of Mombasa to Lake Victoria at Kisumu and finally reached Kampala in 1931.[1] It was built by over forty thousand laborers brought to East Africa from British controlled Punjab, nearly twenty-five hundred of whom died during construction.[2] Of those who survived, about sixty-seven hundred remained in East Africa to find their place in the colonial economy, establishing themselves as traders, running cotton gins, purchasing cash crops, and selling consumer goods to the population of the Uganda Commonwealth, a niche that continued to be dominated by Ugandans of Indian origin until Amin's expulsion orders in 1972.

Traffic on the line declined under the stewardship of the Uganda Railway Corporation (URC). In the context of the political crises of the 1970s and 1980s followed by a period of deep structural adjustment, the URC was unable to invest in the upkeep of the line to compete with growing preferences for road transport. The URC came close to bankruptcy and retrenched its staff massively, from 10,000 at its peak in the 1960s, to 1,164 in 2004, and to only 26 by 2006.[3] The network further deteriorated after it was privatized in 2006, under a widely heralded twenty-five-year concession granted to Rift Valley Railways, who reneged on agreements to invest new capital in the system and saw transit times increase and traffic fall further.[4] Under these conditions of infrastructural abandonment, the tracks became prime venues for popular encroachment. Like similar spaces in cities around the global south, the ambiguous legal status of the tracks and the rail reserves, their proximity to sites of trade and employment, and the context of spiraling housing costs and commute times meant they were valuable sites for squatting. Housing, markets, restaurants, bars, salons, pool halls, cinemas, workshops, factories, and more all took root on this precarious land. URC administrators illegally subdivided and sold off plots of land in the railway reserves to encroachers, and local government officials signed off on the sales. Densely populated but illegal, the rail reserve settlements grew into villages of up to five thousand people but did not benefit from municipal services like piped water, health clinics, schools, or garbage collection. The private national power company, UMEME, eventually connected some of them to its grid in response to the rampant hacking of the power network in these communities.[5] Reliant on polluted wells, the KCC advised residents to get clean water from bottles. Let them drink Evian. NGOs were equally unwilling to engage with these liminal communities, while tenure insecurity and the fear of immanent displacement meant few residents risked investing in their own facilities.

As it transects the city, so too does the railway transect Kampala's waste worlds. The tracks pass by innumerable dumpsites used by local residents who cannot afford private collection and cannot store garbage at home for a month or more waiting for a municipal trash truck to arrive. At Kinawataka, the tracks pass Plastic Recycling Industries, Kampala's largest plastic recycling center, owned by Coca-Cola Bottling, one of the companies responsible for producing much of the plastic waste coursing through Uganda and its waterways. Near here, plastic traders have set up shop along the tracks, buying bottles from itinerant collectors who pick them out of wetlands, dumpsites, drains, bins, and street sides and sell them on at the railside traders' yards just outside the factory gates.

Figure 8. Train tracks run through Ndeeba Market. Photo by research participant
Faith Nakalubi, 2013.

At Namuwongo, on the flat gravel ground adjacent to the rails, the
local government and an NGO have sponsored a similar plastic trading
enterprise as a community business. A few hundred yards down the line,
a concrete bunker is the gazetted garbage dumpsite and collection point
for one of Kampala's most densely packed informal settlements, itself precariously settled along the railway. The tracks pass through the city's main
industrial area, past Mukwano Industries, a plastic factory that produces
the chairs, basins, jerricans, and kitchenware that form the material basis
of everyday domestic life across the city, the remnants of which are a valuable part of plastic traders' dealing. At Katwe, the tracks are full of commuters who walk along the unused line to get directly into the city center
and away from the packed roads of one of the most congested intersections in town.

At Ndeeba, running parallel to a deep drainage channel, the tracks run
alongside markets, houses, and businesses with more dumpsites, scrap deal-
ers' stalls, and plastic traders' yards. Participating in the photo-elicitation

project I conducted, Faith Nakalubi, a young member of a nearby pool and billiards club, photographed the tracks here. Her photos showed a puddle extending between the two tracks and receding to a horizon cross-cut by power lines, the middle ground occupied by pedestrians and vendors making their way along the railway past the bustle of busy shops. She explained:

> This is a water channel near the rail line, but people are running businesses along there. These ones are catching grasshoppers. Children come to catch grasshoppers but can easily fall in the channel. It should be avoided. This is one place that needs improvement. They need to be evicted because it is so squeezed. The market needs to be moved. It is very dirty. I fear for them. They can get sick. There can be a fire. I feel bad when I pass there, but I do not have [any idea] what to do. It looks bad to me. But it is the conditions of the world. They are looking for what to eat.

Faith's reading of this scene speaks to the mundane ways that dirt and displacement, concern and care, development and resignation comingle in ordinary encounters with waste worlds. The conditions along the track clearly need improvement; people are suffering and are at risk of chronic problems of health or catastrophic disasters like fire. Nonetheless, life goes on, as these are the conditions under which people can strive and struggle, earning something to eat and something to take home at the end of the day. Eviction is the easiest form of development to imagine, clearing the space to deal with the entangled risks and indignities it engenders.

Not long after Faith took her photos, evictions did in fact take place along the tracks in Ndeeba, as the railway was reanimated as municipal infrastructure. On July 28, 2014, a public holiday to celebrate Eid al-Fitr, demolitions took place along the rail line, affecting an estimated sixty thousand people.[6] The demolitions were eventually halted by a court order secured by a local women's group, but not before one Ndeeba man died after collapsing at the site of his bulldozed kiosk and decimated livelihood.[7] Nearly four years later, in February 2018, after further rounds of re-encroachment and eviction, the KCCA launched a new commuter rail service, connecting central Kampala to its eastern suburbs. Running parallel to the Kampala-Jinja highway, the train skips some of the city's worst gridlock with a morning and evening service that costs a third of the fare of a minibus taxi and takes just forty-five minutes. Plans are also in the works to redevelop the rail to Kenya to bolster regional integration, ease the movement of people, facilitate trade, and smooth the export of Uganda's landlocked agrarian commodities. But the 2013 proposal to connect Mombasa, Nairobi, Kampala, and Kigali has stalled with delays in anticipated funding from the Chinese government corporation set to operate the line.[8]

The Uganda Railway thus transects Kampala both east to west and past to future. It congeals multiple moments in the city's local history and tracks its changing place in global networks of power, labor, and logistics, from European colonial land grabs to emergent structures of debt and trade oriented toward China, from postcolonial ambitions of modernization to the crises of the 1970s and the failures of neoliberal reforms. Just as it illustrates how the urban population has responded to times of developmental aspiration and chronic crisis, so too does it reveal something about the infrastructure of disposability as it structures and connects diverse waste worlds.

Four processes constitute the condition of disposability throughout the multiple and overlapping waste worlds I've described throughout this book: surplus, embodiment, displacement, and contestation. All are evident and entangled along the Uganda railway. Surplus refers to the processes that create material and social hyperabundance. From the mid-twentieth century, petrocapitalism has been at the root of surplus, creating an overwhelming abundance of plastics, chemicals, and energy that have transformed the geographies and materialities of everyday life around the world and made possible the prevalence and normality of single-use disposable commodities. Raj Patel and Jason Moore have theorized this mode of life as predicated on cheap things—money, nature, work, care, food, energy, and lives—the value of which has been systematically and artificially undermined along the lines of race, gender, class, and nation since Europe began experimenting with plantation capitalism even prior to Columbus.[9] To this list we might add cheap waste.

Regimes of disposability and pollution thrive in part because the costs of the social processes of waste are borne elsewhere, by the people and ecosystems that constitute Away, whose lives and bodies have been made—and are continually remade—as cheap and expendable. In this way, surplus is also a human condition, the state of being exploitable, exterminable, or extraneous within capitalist modernity. Surplus populations are differentially racialized and subject to different forms of violence, from the hyper-exploitation of slavery to the exterminability of genocide to the extraneousness of social abandonment, each rationalized in both the extremes of ideology and the everyday normative unfolding of social life.[10] Along the railway, for instance, we see the colonial use of a racialized laboring class, structurally vital to the construction of empire's infrastructure by being individually expendable. A century on, the rails have become home to another surplus population, urban communities at the margins of contemporary modes of accumulation and excluded from the possibility of even exploitation via waged labor. In contemporary Uganda, the urban poor are considered a developmental anomaly. As a population they are managed simultaneously as threatened (vulnerable to disease) and as a threat (as incubators of disease and disorder). Unemployed urban youth are neither exterminated nor exploited; indeed many express a yearning to be exploited as workers, but instead they are urged to return "to the village" where they are assumed to truly belong. Subject to uplift and reform efforts through rubrics of developmental respectability, they are recognized primarily through a form of care and inclusion predicated on a form of biopolitical racism that justifies the abandonment of the unworthy poor.

In the context of disposability, embodiment refers to the ways that surpluses—of pollutants, wastes, toxins, illnesses, and injuries—unevenly accrue in bodies and are transmitted through some of the most intimate relations of social reproduction: love, nourishment, care, touch, and breath. Because Away is always located somewhere, the hazardous externalities of disposability will remain present in some places and in some lives.[11] When plastic recyclers comb through Kampala's drains to gather the materials they sell to traders along the tracks, for instance, they are exposed to the hazardous wastes and sewage of the city, suffering recurrent and chronic (but often undiagnosed and untreated) illnesses of their stomachs, lungs, and skin. Likewise, waste collectors working for the city embody disposability when they expose themselves, untrained and ill-equipped, to both repetitive stress injuries and horrific accidents on the city's aging fleet of trash trucks.[12] Wastes accrue in and around the precariously built neighborhoods that line the tracks, leading to floods, endemic malaria, and outbreaks of cholera that disproportionately affect the city's poorest residents. But these injuriously embodied conditions of slow violence at home and at work are the conditions of possibility for city life for the global urban population.

Displacement has often been the municipal response to such conditions of embodied disposability around the world, as cities aim to treat injurious urban conditions by simply eradicating them. In this way, disposable surplus populations exposed to the everyday toxicity of waste worlds become triply disposable: the precarious places they have constructed that form the basis of livelihoods become subject to programs of urban clearance lacking adequate plans for re-placement or compensation. Thus, displacement is the effect of new capital investments in profitable enterprises or speculative real estate ventures and can be the outcome of interventions imagined as forms of care, development, and humanitarian assistance.[13] In these cases, "for their own good," people's homes are destroyed and livelihoods criminalized. At the most material level, displacement generates wastes when bulldozers turn shops and homes to rubble. Rhetorically, this destruction proceeds through the construction of such spaces as already-wastelands, valueless problem spaces that must be valorized through their integration into the circuits of capitalist production.[14] Cleaning is world making, as Mary Douglas argued half a century ago. Urban cleaning worlds in ways simultaneously desirable and inequitable, offering aspirational horizons that often affectively appeal to the very people who will be displaced to create them.[15] Both versions of displacement, speculative and humanitarian, that took place have shaped the waste worlds of the Uganda Railway. The construction of an upmarket housing estate at Nakawa displaced long-term tenants there who found a new place to live in shacks built in the rail reserve, and they were later displaced again when the train was redeveloped by the KCCA. Elsewhere, evictions also took place in the name of clearing obstacles to urban development, simultaneously victims and causes of environmental injustice.

But these processes of disposability do not unfold smoothly or without contestation. Somewhere between the dramatic, persistent, and romanticized notion of resistance and the grim everyday accommodation to chronic crises that is resilience, contestation refers to the intermittent, often reactive, and contingent ways that people protest, slow, respond to, and demand alternatives to displacement. Small-scale local protests and alliances between residents; market traders; and other precarious businesses, politicians, and sympathetic journalists, for example, were able to slow the evictions along the rail lines, creating a pause in displacement and urging the municipal government to provide alternatives to those whose homes and shops were facing bulldozers. Such forms of contestation are typical responses around the world when forms of mundane urban encroachment, squatting, and extra-legal inhabitation—processes through which the everyday life of vast swaths of the urban population are constituted and through which disposable lives claim and create their place in the city—are threatened and undermined. While contestation over displacement can bring temporary victories, it often leaves intact other processes of disposability. It is harder to identify and protest, let alone to undo, the slower processes of disposability as surplus and embodiment. Surplus, embodiment, displacement, and contestation thus form disposability's infrastructure, the socio-material processes through which this form of production, valuation, and destruction takes shape. But waste worlds are densely inhabited, and disposable lives are, nonetheless, lived.

Notes

INTRODUCTION

1. Parnell and Pieterse, *Africa's Urban Revolution*.
2. Roy, "Who's Afraid of Postcolonial Theory?"; Myers, *Rethinking Urbanism*.
3. Douglas, *Purity and Danger*.
4. Liboiron, "'Matter Out of Place.'"
5. Alaimo, *Exposed*.
6. Bennett, *Vibrant Matter*, 4–6.
7. Bennett, 99.
8. Gregson and Crang, "Materiality and Waste."
9. Goldstein, "Waste"; Lucas, "Disposability and Dispossession."
10. Rogers, *Gone Tomorrow*; Gabrys, Hawkins, and Michael, *Accumulation*.
11. Royte, *Garbage Land*; Hoy, *Chasing Dirt*.
12. Strasser, *Waste and Want*, 162–64.
13. Higgins, "Essential Workers."
14. Reno, *Waste Away*.
15. Packard, *The Wastemakers*.
16. Scanlan, *On Trash*; Hawkins, *Ethics of Waste*; Kennedy, *Ontology of Trash*.
17. Rathje and Cullen, *Rubbish!*
18. Doron and Jeffrey, *Waste of a Nation*; Knowles, *Flip-Flop*; Nguyen, *Waste and Wealth*.

19. Gregson and Crang, "Waste to Resource"; Minter, *Junkyard Planet*; Gill, *Poverty and Plastic*.

20. Gregson, Metcalfe, and Crewe, "Object Maintenance and Repair"; Jackson, "Rethinking Repair."

21. Barnard, *Freegans*; Giles, "Anatomy of a Dumpster."

22. Stamatopoulou-Robbins, *Waste Siege*, 170.

23. *Discard studies* is Max Liboiron's term for the interdisciplinary field of research that examines not just waste flows themselves but also assumptions about where waste is produced, by whom, and how to see it. Importantly, while the term *waste* tends to focus attention on individual objects and practices, the concept of discard, they propose, highlights the systems that make disposability possible. Liboiron, "Why Discard Studies?"

24. Reno, *Waste Away*; Nagel, *Picking Up*; Melosi, *Garbage in the Cities*; Liboiron, *Pollution*.

25. Bullard, *Dumping in Dixie*; Checker, *Polluted Promises*; Lerner, *Sacrifice Zones*; Murphy, "Chemical Infrastructures"; Pulido, "Critical Review"; T. Taylor, *Toxic Communities*; Vasudevan, "Intimate Inventory."

26. Voyles, *Wastelanding*; Stamatopoulou-Robbins, *Waste Siege*.

27. Davis, "Toxic Progeny."

28. Alexander and Reno, *Economies of Recycling*; Gregson and Crang, "Waste to Resource."

29. For a sampling of these theoretical projects, see *Histories of Violence*, an online project featuring videos of prominent contemporary theorists including Gil Anidjar, Etienne Balibar, Simon Critchley, Lewis Gordon, Saskia Sassen, and Slavoj Žižek reflecting on the theme "disposable life." Evans, "Disposable Life."

30. Evans and Giroux, *Disposable Futures*, 45.

31. Bauman, *Wasted Lives*, 12.

32. Doherty and Brown, "Labor Laid Waste," 5–6.

33. Millar, *Reclaiming the Discarded*.

34. Millar, "Garbage as Racialization."

35. These numbers are based on World Bank, "Uganda Economic Update," 47, and Kampala Capital City Authority, "Kampala Physical Development Plan," 102.

36. McFarlane and Silver, "Navigating the City."

37. Desai, *We Are the Poors*; Makhulu, *Making Freedom*; Pellow, *Landlords and Lodgers*; Ross, *Raw Life*; Nunzio, *Act of Living*.

38. J. Ferguson, *Give a Man a Fish*; Watson, "Seeing from the South."

39. Bhan, "Bold Plan."

40. Pieterse, "Grasping the Unknowable."

41. Gandy, "Learning from Lagos."

42. The blog *Africa Is a Country* has published numerous critiques of this discourse. See Tveit, "The Afropolitan Must Go"; Dabiri, "Why I'm Not an Afropolitan"; Kibala Bauer, "Beyond the Western Gaze."

43. Mains, *Under Construction*; Tsing, "Inside the Economy"; Watson, "African Urban Fantasies."

44. This is the domain of a rich body of historical and ethnographic scholarship exemplified recently by, among others, Quayson, *Oxford Street*; Makhulu, *Making Freedom*; Mains, *Under Construction*; Morton, *Age of Concrete*; Boeck and Baloji, *Suturing the City*.

45. Ghertner, *Rule by Aesthetics*; Baviskar and Ray, *Elite and Everyman*.

46. Robinson, *Ordinary Cities*.

47. Roy and Crane, *Territories of Poverty*; Roy et al., *Encountering Poverty*.

48. Li, *Land's End*, 3.

49. Grainger and Geary, "New Forests Company"; *The Guardian*, "Ugandan Farmers"; National Association of Professional Environmentalists, *Land Grabbing*; Transparency International Uganda, *Up against Giants*; Akaki, *Mabira Forest Giveaway*.

50. Lindell, *Africa's Informal Workers*; Kinyanjui, *Women and the Informal Economy*; Rizzo, *Taken for a Ride*.

51. Rizzo, Kilama, and Wuyts, "Invisibility of Wage Employment."

52. Donovan, "Colonizing the Future"; Roy, *Poverty Capital*.

53. Davis, *Planet of Slums*; Meagher, "Cannibalizing the Informal Economy"; Meagher and Lindell, "African Informal Economies"; Hickey and du Toit, "Adverse Incorporation"; K. Taylor, "Predatory Inclusion."

54. Gidwani, "Work of Waste"; Pereira da Silva, "Follow the Bottle."

55. Patel and Moore, *History of the World*.

56. Stamatopoulou-Robbins, *Waste Siege*, 108–10.

57. *The Observer*, "Museveni's Four-Acre Prosperity Formula."

58. Museveni, *Sowing the Mustard Seed*.

59. Cited in Ranciere, *Philosopher and His Poor*, 95.

60. Marx, *Eighteenth Brumaire*; see also Stallybrass, "Marx and Heterogeneity."

61. Fanon, *Wretched of the Earth*, 103.

62. Fanon, 103.

63. Reno, *Waste Away*.

64. Gabrys, "Sink."

65. Livingston, *Self-Devouring Growth*.

66. Kirby, "Slow Burn."

67. *Seattle Times*, "Drugs Found."

68. Murphy, "Chemical Regimes."

69. Nixon, *Slow Violence*, 6.

70. Hawkins, *Ethics of Waste*; Žižek, "Ecology."

71. *New Vision*, "Four Killed in Kampala Floods."

72. *Daily Monitor*, "Government Tackles Flooding"; *New Vision*, "Museveni Orders."

73. Reuters, "Three Die."

74. Mukiibi, "Policies on the Supply of Housing"; Nuwagaba, "Impact of Macro-Adjustment Programmes"; Appelblad Fredby and Nillson, "All for Some"; Katusiimeh, Mol, and Burger, "Operations and Effectiveness."

75. Winner, "Do Artifacts Have Politics?"

76. Althusser, *Reproduction of Capitalism*; Mann, "Autonomous Power"; Scott, *Domination and the Arts*. While this literature is now too vast to summarize here, two recent collections gathering a cross-section of researchers in the field serve as valuable points of entry: Anand, Gupta, and Appel, *Promise of Infrastructure*; Hetherington, *Infrastructure, Environment, and Life*.

77. Larkin, "Politics and Poetics"; Melly, *Bottleneck*.

78. Rodgers and O'Neill, "Infrastructural Violence."

79. De Leon, *Land of Open Graves*; Gilmore, *Golden Gulag*; Doherty, "Big Bellies."

80. Rao, "How to Read a Bomb"; Stamatopoulou-Robbins, *Waste Siege*; S. Graham, "Lessons in Urbicide."

81. Winner, "Do Artifacts Have Politics"; Bullard, *Dumping in Dixie*.

82. Doherty, "Life (and Limb)."

83. Povinelli, *Economies of Abandonment*, 131–62.

84. Larkin, "Politics and Poetics," 329.

85. Star, "Ethnography of Infrastructure."

86. J. Ferguson, *Global Shadows*; Kimari and Ernstson, "Imperial Remains."

87. Melly, *Bottleneck*; Mains, *Under Construction*; Larkin, "Promising Forms."

88. Mitchell, *Rule of Experts*; Bloom, Miescher, and Manuh, *Modernization as Spectacle*; Schindler and Kanai, "Getting the Territory Right."

89. Leonard, *Life in the Time of Oil*.

90. Von Schnitzler, *Democracy's Infrastructure*; Chalfin, "Waste Work"; Samson, "Producing Privatization"; Rizzo, *Taken for a Ride*; Mains, "Blackouts and Progress."

91. Star, "Ethnography of Infrastructure."

92. Lawhon et al., "Heterogeneous Infrastructure Configurations."

93. Agrawal, *Environmentality*; Argyrou, "Keep Cyprus Clean."

94. In the 2006–2007 fiscal year, 22 percent of the KCC's USh55 billion budget (US$22 million; USh denotes Ugandan shillings)—USh2.4 billion (approximately US$965,000)—was spent on waste collection and disposal. In the 2012–2013 KCCA annual budget, USh17.33 billion (about US$6.9 million) was allocated to waste management, 10.6 percent of the total municipal budget of USh162.58 billion (about US$65 million).

95. Bjerkli, "Governance on the Ground"; Chikarmane, *Integrating Waste Pickers*; Fergutz, "Developing Urban Waste Management."

CHAPTER 1. ACCUMULATIONS OF AUTHORITY

1. *The Independent*, "Jennifer Musisi's Kampala."
2. Ezek. 6:6
3. Ezek. 33:28
4. J. S. Musisi, "Statement on the Issues."
5. Oliver, "Royal Tombs"; Ray, "Royal Shrines."
6. D. Apter, *Political Kingdom in Uganda*.
7. Neil Kodesh argues that accounts of Ganda history have overemphasized the singularity of the kabaka's authority and ignored the role of clan-based spirit mediums, whose authority as public healers was diffuse and decentralizing. He attributes this to an uncritical acceptance of early accounts written by Apollo Kagwa, a katikiro and chief who, in addition to his own books *The Kings of Buganda* and *The Customs of the Baganda*, was essentially the coauthor of missionary John Roscoe's ethnography *The Baganda*. Kagwa, Kodesh argues, was politically motivated to portray authority in the kingdom as centralized in the kabaka in order to consolidate the power and landholdings of chiefs, such as himself, whose ultimate authority derived from the kabaka. Kodesh, *Beyond the Royal Gaze*.
8. Low, "Religion and Society"; Hanson, *Landed Obligation*; Reid, *Political Power*; Tiberondwa, *Missionary Teachers*.
9. Kigongo and Reid, "Local Communities."
10. Karlström, "Imagining Democracy" and "Modernity and Its Aspirants."
11. The seat of British administration was not actually Kampala, however, but Entebbe (the Luganda word for *seat*), a garden city twenty-five miles to the south, where, at the time of writing, President Museveni has his primary official residence, State House Entebbe, a short drive from Uganda's primary airport, a staging ground for United Nations operations across East Africa.
12. Southall and Gutkind, *Townsmen in the Making*; Parkin, *Neighbors and Nationals*.
13. Omolo-Okalebo et al., "Planning of Kampala City."
14. Gutschow, "Modern Planning" and "Das Neue Afrika."
15. Stoler, *Carnal Knowledge and Imperial Power*.
16. Summers, "Intimate Colonialism"; Kyomuhendo and McIntosh, *Women, Work and Domestic Virtue*.
17. Namara, "Invisible Workers."
18. Bashford, *Imperial Hygiene*.
19. Quoted in Gutkind, *Royal Capital of Buganda*, 122–24.

20. Vaughan, *Curing Their Ills.*

21. Gutkind, *Royal Capital of Buganda*, 122–40.

22. Mirams, *Kampala*, 121.

23. Mirams, 121.

24. Mbembe, "Edge of the World," 266.

25. D. Apter, *Political Kingdom in Uganda*, 29.

26. Nuwagaba, "Dualism in Kampala," 151; Nkurunziza, "Two States, One City?"

27. Myers, *Verandahs of Power*; Burton, *African Underclass*; White, *Comforts of Home.*

28. Republic of Uganda, "Kampala Development Plan."

29. Kasfir, "State, Magendo, and Class"; Mamdani, "Ugandan Asian Expulsion"; Twaddle, *Expulsion of a Minority*; Obbo, *African Women.*

30. Hansen and Twaddle, *Changing Uganda*; Barr, Fafchamps, and Owens, "Governance of Non-Governmental Organizations."

31. John van Nostrand Associates, *Kampala Urban Study.*

32. Goodfellow, "'Bastard Child'"; Mukwaya, Sengendo, and Lwasa, "Urban Development Transitions."

33. Vermeiren et al., "Urban Growth of Kampala."

34. Vermeiren et al., 205.

35. Mukwaya, Sengendo, and Lwasa, "Urban Development Transitions"; Sengendo et al., "Informal Sector"; Nkurunziza, "Informal Mechanisms."

36. Republic of Uganda, "Free Zones Act"; Brautigam and Ziaoyang, "African Shenzhen."

37. Obbo, "Women, Children and a Living Wage"; Nuwagaba, "Macro-Adjustment Programmes."

38. Whyte et al., "Therapeutic Citizenship," 143.

39. Whyte et al., 143.

40. Mwenda, "Personalizing Power"; Tripp, *Museveni's Uganda*; Carbone, *No Party Democracy*; Rubongoya, *Regime Hegemony.*

CHAPTER 2. TEAR GAS AND TRASH TRUCKS

1. Philipps and Kagoro, "Metastable City"; Kagoro, "Competitive Authoritarianism."

2. Tripp, *Museveni's Uganda*; Sjogren, *Militarism and Technocratic Governance.*

3. Green, "Patronage, District Creation, and Reform."

4. Oloka-Onyango, "New-Breed Leadership."

5. Abrahamsen and Bareebe, "Uganda's 2016 Elections."

6. Mwenda and Tangri, "Patronage Politics"; Mwenda, "Personalizing Power."

7. F. Cooper, *Africa in the World*, 30.

8. Sociologist Michael Mann defines infrastructural power, in contrast to despotic power, as "the capacity of the state actually to penetrate civil society, and to implement logistically political decisions through the realm." Within Mann's framework, the establishment of the KCCA in a state with high despotic power represents a transition from an imperial (low-infrastructural power) to authoritarian (high infrastructural power) ideal type. Mann, "Autonomous Power," 113.

9. Doherty, "Maintenance Space."

10. Chu, "When Infrastructures Attack."

11. A. Alexander, "Disciplining Method"; Semboja and Therkildsen, *Service Provision*.

12. Resnick, *Urban Poverty and Party Populism*.

13. Goodfellow, *Bastard Child*; Gore and Muwanga, "Decentralization Is Dead"; Lambright, "Opposition Politics."

14. I encountered these views of the city's history among Kampala residents aligned with either side of the conflict as well as in the rhetoric of politicians and political analysts/editorialists in the national print, radio, and television media.

15. Katusiimeh and Mol, "Environmental Legacies."

16. Baral, "Chicken and the Egg"; Green, "Decentralization and Development."

17. Uganda Media Center, "President Says."

18. *The Observer*, "Musisi Wants Lukwago"; *The Observer*, "KCCA at 1 Year."

19. Young, "Protection to Repression."

20. *The Independent*, "Kampala's Lord Mayor Lukwago."

21. Izama and Wilkerson, "Museveni's Triumph and Weakness."

22. Branch and Mampilly, *Africa Uprising*.

23. Human Rights Watch, "Launch Independent Inquiry."

24. Izama and Echwalu, "Season of Dissent."

25. *The Guardian*, "Opposition Leader Temporarily Blinded."

26. *Daily Monitor*, "I Will Continue Walking."

27. Goodfellow, "Institutionalisation of 'Noise'"; Philipps and Kagoro, "Metastable City."

28. Izama, "Museveni Still the West's Man"

29. Amnesty International, "Stifling Dissent."

30. Goodfellow, "Legal Manoeuvres and Violence."

31. Hundle, "Insecurities of Expulsion"; Moore, "What the Miniskirt Reveals."

32. Speaker of Parliament, Rebecca Kadaga, voiced this rhetoric most loudly, accusing Canada's foreign affairs minister of intimidation and failing to respect Uganda's sovereignty at an international meeting in Quebec. *Daily Monitor*, "Speaker Kadaga Promises."

33. Decker, *In Idi Amin's Shadow*.

34. Human Rights Watch, "Stop Harassing the Media"; Reporters without Borders, "Two Kampala Newspapers."

35. Baral, "Chicken and the Egg."

36. From March 2013, Lukwago was involved in a protracted legal struggle with the KCCA who sought to remove him from office. Prior to his November 25, 2013, impeachment, the High Court issued a court injunction against the impeachment vote. KCCA officials proceeded anyway, physically blocking Lukwago's lawyer from delivering the court order to the impeachment hearing, where they voted 29–3 to impeach. This impeachment was annulled days later, and Lukwago was reinstated, but the attorney general appealed this decision. In March 2014, a judge barred Lukwago from carrying out his duties as mayor pending the attorney general's appeal. While the case continued, Lukwago retained the title lord mayor and sought full reinstatement to the role he characterized as oversight, audit, and accountability. He was reelected in the 2016 and 2021 elections.

37. *New Vision*, "Kampala Is Better Now."

38. *Daily Monitor*, "New KCCA Law"; *The Independent*, "Museveni Killed Anti-Lukwago Bill."

39. This dynamic is not unique to Kampala. As Stamatopoulou-Robbins illustrates, in Palestine, waste management is a critical municipal function that can afford contested states the chance to perform their capabilities and hone the professional skills of staffers charged with urban management. Stamatopoulou-Robbins, *Waste Siege*, 41.

40. Young, "Protection to Repression," 725.

41. *Le Monde Diplomatique*, "Cheap Help from Uganda." I first encountered this fact as a rumor and posed it as a question in an interview with the KCCA official in charge of the enforcement officer program. He confirmed—very casually, without regarding the issue as controversial—that many of his recruits had experience with private security contractors in Iraq, although he could not provide a percentage.

42. Baral, "Chicken and the Egg."

43. In 2017, running as an independent candidate under his legal name, Robert Kyagulanyi Ssentamu, Bobi Wine was elected MP for Kyadondo East, a northern Kampala constituency that includes Kamwokya slum, a neighborhood that is home to his recording studio and is where he led a Kampala City Yange cleanup event in 2012. He contested the 2021 presidential election against Museveni, suffering the same forms of violent state repression that Besigye had previously endured before losing what he contends was a fraudulent vote.

44. Hansen and Vaa, *Reconsidering Informality*; Myers, *Rethinking Urbanism*; Parnell and Pieterse, *Africa's Urban Revolution*.

45. *Daily Monitor*, "Jennifer Musisi"; *New Vision*, "Face to Face"; Radio Netherlands Worldwide, "People Threaten to Kill Me"; *Daily Monitor*, "Fascism, City, Country."

46. Bin-It Services Limited v. Kampala Capital City Authority & Another, 593 (High Court of Uganda at Kampala, Civil Division, April 14, 2020).

47. This was part of a comprehensive cleanup of the municipal administration itself via a complete staff overhaul. Every KCC worker had their contract terminated and had to reapply for a position in the new authority.

48. *New Vision*, "Kampala Is Better Now."

49. J. Ferguson, *Anti-politics Machine*; Li, *Will to Improve*.

50. Young, "Protection to Repression," 718–20.

51. Woolgar, "Configuring the User."

52. S. T. Brown, "Kampala's Sanitary Regime."

53. *Daily Monitor*, "Shs485b Key Development Plans"; Republic of Uganda, "Money Audit Report"; Kampala Capital City Authority, *Strategic Plan 2014/15–2018/19*.

54. Reno, "Waste and Waste Management."

55. Jackson, "Rethinking Repair"; Mattern, "Maintenance and Care"; Graziano and Trogal, "Repair Matters."

56. iFixit, "Self-Repair Manifesto."

57. Hetherington, "Secondhandedness"; Gregson, Metcalfe, and Crewe, "Object Maintenance and Repair"; Reno, "Your Trash."

58. Vinsel and Russell, "Hail the Maintainers."

CHAPTER 3. DESTRUCTIVE CREATION

1. *New Vision*, "KCCA Demolishes Kiosks"; *Daily Monitor*, "New Taxi Park Renovation."

2. Ezek. 33:28 and 33:33.

3. Schumpeter, *Capitalism, Socialism and Democracy*; Gordillo, *Rubble*, 80.

4. Gordillo, *Rubble*, 81.

5. Harms, *Luxury and Rubble*, 7.

6. Gordillo, *Rubble*, 80.

7. Reinert, "Sacrifice."

8. Lerner, *Sacrifice Zones*; Kuletz, *Tainted Desert*.

9. This environmental and rhetorical process continues to underpin the construction of American energy infrastructure and the management of its toxic afterlives. Voyles, *Wastelanding*; Cram, "Wild and Scenic Wasteland."

10. *New Vision*, "Nakawa-Naguru Tenants Evicted."

11. *Daily Monitor*, "Nakawa-Naguru Families Stranded."

12. Frearson, "David Adjaye."

13. Boeck, "Spectral Kinshasa," 323.

14. *Daily Monitor*, "1,700 New Housing Units"; *Daily Monitor*, "IGG Denies"; *Daily Monitor*, "Ex-PS Explains."

15. *RedPepper*, "Govt"; *The Independent*, "Remembering the KCCA Evictions."

16. *Daily Monitor*, "Railway Eviction Case Adjourned"; *New Vision*, "Rubaga Leaders"; *Daily Monitor*, "KCCA Tests Passenger Train"; *The Independent*, "Commuter Train Service"; *New Vision*, "KCCA to Introduce Railway Service."

17. Gupta, "Suspension."

18. Stamatopoulou-Robbins, "Failure to Build," 8–9.

CHAPTER 4. SELFIES OF THE STATE

1. The content shared across these platforms is largely the same. I chose to focus on the KCCA's Facebook page because it was the livelier venue for public commentary in the period of research and provided a more discretely organized space for public commentary than Instagram or Twitter.

2. A. Apter, *Pan-African Nation*; Boeck, "Inhabiting Ocular Ground"; Melly, *Bottleneck*.

3. Joyce, *Rule of Freedom*.

4. Otter, "Making Liberalism Durable" and "Cleansing and Clarifying"; Barry, Osborne, and Rose, *Foucault and Political Reason*.

5. Kotef, *Movement and the Ordering of Freedom*.

6. S. Graham, *Disrupted Cities*.

7. Heidegger, *Being and Time*. Elaborating on this idea, Sara Ahmed writes, "Failure, which is about the loss of the capacity to perform an action for which the object was intended is not a property of an object (though it tends to be attributed in this way and there is no doubt that things can go wrong), but rather of the failure of an object to extend a body, which we can define in terms of the extension of bodily capacities to perform actions." Ahmed, *Queer Phenomenology*, 49.

8. Lakoff and Collier, "Infrastructure and Event"; Masco, "'Survival Is Your Business'"; Page, *City's End*.

9. Mukherjee, *Radiant Infrastructures*.

10. Prakash, *Another Reason*; Mitchell, *Rule of Experts*; A. Apter, *Pan-African Nation*; Street, *Biomedicine in an Unstable Place*; Bloom, Miescher, and Manuh, *Modernization as Spectacle*.

11. Kampala Capital City Authority, *Strategic Plan 2014/15–2018/19*.

12. Nye, *American Technological Sublime*.

13. McFarlane and Silver, "Navigating the City."

14. Taussig, *Defacement*; Mbembe, *On the Postcolony*.

15. Kampala Capital City Authority, "Shut Down of Operations."

16. *Daily Monitor*, "Musisi Ordered to Open KCCA Offices."

17. *New Vision*, "KCCA Employees Back to Work."

18. *New Vision*, "Who Is Responsible?"

19. BBC, "Protesters Fish in Uganda Pothole."

20. Mains, *Under Construction*, 111.

21. Scott, *Seeing Like a State*.

22. Pierce, "Looking Like a State."

23. J. S. Musisi, "Statement on the Issues."

CHAPTER 5. PARA-SITES

1. *Loader* is the word used to name municipal and private workers who load garbage trucks. Self-employed plastic and other scrap collectors working at the landfill and in town use a variety of names for their work. Some call it, simply, *kasasiro* (garbage); others use the label *waste picker, waste collector,* or *salvager*. For the sake of clarity, I use the term *salvager* to refer to the work of sorting and gathering salable materials (be it at the landfill or in town) and the term *waste collector* to refer to self-employed people who gather and dispose of waste in residential and commercial areas. The term *waste picker* was most widely used by NGOs and municipal workers. It did not distinguish between these two types of work and could also encompass loaders. No one involved in these jobs used the term *scavenger*, although it was used in policy documents and in the press.

2. Stamatopoulou-Robbins, *Waste Siege*, 213.

3. Doherty and Brown, "Labor Laid Waste."

4. Muir and Gupta, "Rethinking the Anthropology of Corruption."

5. Sedgwick, *Touching Feeling*, 8.

6. Gidwani, *Capital, Interrupted*, 147.

7. Pratt, *Imperial Eyes*, 4.

8. Kasozi, *Social Origins of Violence*.

9. G. Cole, "Historical Backlash."

10. Raffles, "Jews, Lice, and History"; Mamdani, *When Victims Become Killers*.

11. Hird, *Origins of Sociable Life*; Paxson and Helmreich, "Perils and Promises."

12. Benezra, DeStefano, and Gordon, "Anthropology of Microbes."

13. Lorimer, "Parasites, Ghosts and Mutualists."

14. Brunner et al., "Experimental Evidence."

15. Lorimer, "Parasites, Ghosts and Mutualists."

16. Pratt, *Imperial Eyes*, 6.

17. Mitchell, *Rule of Experts*; Roy, *Poverty Capital*.

18. Scott, *Domination and the Arts*, 183–84.

19. Chattopadhyay, *Unlearning the City*, 51–54.

20. Graeber, *Direct Action*, 203.

21. Bayat, *Life as Politics*; Benjamin, "Occupancy Urbanism."

22. Gidwani, "Work of Waste," 576.

CHAPTER 6. LEGALIZING WASTE

1. Cointreau, "Sanitary Landfill Design"; Cointreau, "Occupational and Environmental Health"; Hoornweg and Bhaba-Tata, *What a Waste*; Schubeler, Wehrle, and Christen, "Conceptual Framework."

2. Mains, *Under Construction*, 14–17.

3. Douglas, *Purity and Danger*.

4. Hawkins, *Ethics of Waste*.

5. Laporte, *History of Shit*.

6. Kampala City Council, "(Solid Waste Management) Ordinance," 5.

7. Golooba-Mutebi, "Devolution and Outsourcing of Municipal Services in Kampala City, Uganda"; Kampala City Council, "Strategy to Improve Solid Waste Management in Kampala City (1999/2002)."

8. On the politics of privatization and outsourcing of municipal waste management in Africa, see Bjerkli, "Governance on the Ground"; Fahmi, "Impact of Privatization"; Myers, *Disposable Cities*; Oteng-Ababio, Melara Arguello, and Gabbay, "Solid Waste Management"; Samson, "Producing Privatization."

9. German Technical Cooperation Agency, "Solid Waste Disposal."

10. Mugaga, "Public-Private Sector Approach"; Katusiimeh, Mol, and Burger, "Operations and Effectiveness"; Katusiimeh, Burger, and Mol, "Informal Waste Collection"; Gore and Muwanga, "Decentralization Is Dead."

11. Republic of Uganda, *Money Audit Report*.

12. Golooba-Mutebi, "Devolution and Outsourcing."

13. Katusiimeh and Mol, "Environmental Legacies."

14. *New Vision*, "Kampala a Mountain of Garbage."

15. *New Vision*, "One Year on."

16. Graham and Marvin, *Splintering Urbanism*.

17. Kooy and Bakker, "Splintered Networks."

18. Mamdani, *Politics and Class Formation in Uganda*; Nuwagaba, "Dualism in Kampala"; Omolo-Okalebo et al., "Planning of Kampala City"; Appelblad Fredby and Nillson, "All for Some"; S. T. Brown, "Planning Kampala."

19. Reno, *Waste Away*, 99–101.

20. Agrawal, *Environmentality*, 164.

21. Holmes and Marcus, "Fast Capitalism," 35.

22. Lindell and Appelblad, "Disabling Governance"; Gombay, "Eating Cities."

23. Katusiimeh, Mol, and Burger, "Operations and Effectiveness."

24. Serres, *Parasite*, 12–13, 79. Despite the post-structuralist influence on, and reception of, his thought, Serres's primary engagement with systems theory and his concern to show the ways that systems work through failure mean that he produces a kind of post-structural functionalism. Although he does not reduce parasites to their role in the reproduction of systems, this is his primary

analytic concern. Moreover, the formalism of his approach privileges sameness and abstraction instead of understanding the different roles para-sites play in different systems.

25. Scott, *Seeing Like a State*.

26. Serres, *Parasite*, 13.

27. Tsing, *Friction*.

CHAPTER 7. SINK AND SPILL

1. Prior to the opening of Kiteezi Landfill, garbage was dumped at open dumpsites around Kampala, typically located at the edges of the city's many wetlands. On the composition of waste deposited at Kiteezi, see Komakech et al., "Characterization of Municipal Waste in Kampala, Uganda."

2. Gabrys, "Sink."

3. Bize, "Right to the Remainder," building on the argument made in Tsing, *Friction*.

4. Solomon, "Ghetto Is a Gold Mine"; Gidwani, "Work of Waste."

5. Pereira da Silva, "Follow the Bottle."

6. Schulz and Lora-Wainwright, "In the Name of Circularity."

7. Liboiron, "Recycling as a Crisis of Meaning"; MacBride, *Recycling Reconsidered*.

8. Millar, *Reclaiming the Discarded*, 71.

9. Reno, *Waste Away*, 43–35.

10. Doherty, "Motorcycle Taxis."

11. Talbott, "A Green Army Is Ready."

12. Kampala City Council, "(Solid Waste Management) Ordinance," 14.

13. The avian-focused environmental NGO Nature Uganda conducts an annual vulture count to track the numbers of marabou storks and other vulture species in the city. The birds, concentrated at Kiteezi Landfill, on the campus of Makerere University, near the city's abattoirs, and in the diffuse wetland dumpsites, register the intersections of the waste stream's para-sites and the city's green space, places where both suitable trees for nesting and readily available food sources can be found.

14. Haraway, *When Species Meet*; Nagy and Johnson, *Trash Animals*; Doherty, "Filthy Flourishing."

15. Kampala City Council, "Strategy to Improve," 4.

16. Anand, *Hydraulic City*.

17. Alexander and Reno, *Economies of Recycling*.

18. Nguyen, *Waste and Wealth*; Millar, "Precarious Present"; Gregson et al., "Following Things of Rubbish Value."

CHAPTER 8. ASSEMBLING THE WASTE STREAM

1. Fredericks, *Garbage Citizenship*, 94–96.
2. Law, *Organizing Modernity*.
3. Scott, *Seeing Like a State*.
4. National Environmental Management Authority, "Draft Guidelines," 10.
5. National Environmental Management Authority, 62.
6. Niringiye and Douglason, "Determinants of Willingness," 121.
7. National Environmental Management Authority, "Draft Guidelines," 22.
8. Baudouin et al., "Between Neglect and Control"; Demaria and Schindler, "Contesting Urban Metabolism"; Fergutz, "Developing Urban Waste Management."
9. Miraftab, "Neoliberalism and Casualization"; Samson, "Rescaling the State."
10. Fredericks, *Garbage Citizenship*, 98–101.
11. Anand, *Hydraulic City*.

CHAPTER 9. EMBODIED DISPLACEMENT

1. Douglas, *Purity and Danger*, 2.
2. Bauman, *Wasted Lives*.
3. Serres, *Malfeasance*, 3.
4. Serres, 2.
5. S. Brown, "Michel Serres."
6. Whitson, "Negotiating Place and Value"; Fergutz, "Developing Urban Waste Management"; Baudouin et al., "Between Neglect and Control"; Rockson et al., "Activities of Scavengers"; Chikarmane, *Integrating Waste Pickers*.
7. Ghertner, *Rule by Aesthetics*.
8. Fanon, *Wretched of the Earth*, 30–31.
9. Murphy, "Chemical Infrastructures."
10. Nixon, *Slow Violence*, 6.
11. D. Graham, "History of *Helicobacter Pylori*."
12. Hestvik et al., "Prevalence of *Helicobacter Pylori*."
13. Alvarado-Esquivel, "*Helicobacter Pylori* Infection."
14. Livingston, *Improvising Medicine*.
15. An environmental analysis of the landfill found that the nearby wetland was filtering leachate and absorbing pollutants but that conflicting land-uses make this filtration unsustainable and hazardous. Mwiganga and Kasiime, "Impact of Mpererwe Landfill." A study of lead exposure linked to the transition to unleaded gasoline suggests that Kiteezi Landfill, via groundwater contamination, is a source of lead exposure for children living nearby. Graber et al., "Childhood Lead Exposure."
16. Mbembe and Roitman, "Figures of the Subject."
17. Rodgers and O'Neill, "Infrastructural Violence," 404.

CHAPTER 10. FROM NATIVES TO LOCALS

1. Moore, "Idle Women"; Wyrod, *AIDS and Masculinity.*
2. S. T. Brown, "Planning Kampala," 84.
3. Newell, *Histories of Dirt.*
4. Pierre, *Predicament of Blackness.*
5. Adichie, "Interview with *'Americanah'* Author."
6. Hurston, "How It Feels," 154.
7. Robinson, *Black Marxism.*
8. Ndlovu-Gatsheni, "Coloniality of Power"; Pailey, "De-Centering the 'White Gaze.'"
9. Moore, "Postures of Empowerment."
10. Dill, *Fixing the African State*; Lie, *Developmentality.*
11. Bernal and Grewal, *Theorizing NGOs,* 10.
12. In the United States, community development policies and community participation directives have played a similar role in channeling radical politics and efforts at self-determination toward bureaucratic state programs to manage racialized poverty. These programs were part of a global Cold War policy concern with preempting insurrection and otherwise mitigating the political risks of extreme inequality. Goldstein, *Poverty in Common*; Roy, Schrader, and Crane, "Anti-Poverty Hoax."
13. Dill, *Fixing the African State.*
14. Miraftab, "Neoliberalism and Casualization."
15. Fredericks, *Garbage Citizenship,* 99.
16. Doherty, "Capitalizing Community."
17. Bornstein and Sharma, "Righteous and the Rightful."
18. Ferguson and Gupta, "Spatializing States."
19. Escobar, *Encountering Development*; Pailey, "De-Centering the 'White Gaze'"; Ndlovu-Gatsheni, "Coloniality of Power"; Crewe and Fernando, "Elephant in the Room"; Kothari, "Agenda for Thinking about 'Race.'"
20. Moore, "Postures of Empowerment."
21. Loftsdóttir, "Invisible Colour."
22. Pierre, "Structure, Project, Process," 215–17.
23. Pierre, "Racial Vernaculars of Development."

CHAPTER 11. INFRASTRUCTURES OF FEELING

1. Omi and Winant, *Racial Formation,* 71.
2. Appelblad Fredby and Nillson, "All for Some"; S. T. Brown, "Planning Kampala"; Gutschow, "Das Neue Afrika"; Monteith, "Markets and Monarchs"; Omolo-Okalebo et al., "Planning of Kampala City."

3. Pierre, *Predicament of Blackness*, 42.

4. Kimari and Ernstson, "Imperial Remains," 841.

5. Williams, *Marxism and Literature*, 133.

6. Williams, *Long Revolution*, 65.

7. Green-Simms, *Postcolonial Automobility*, 25; see also Mains, *Under Construction*, 93–96; Melly, *Bottleneck*.

8. Massumi, "Autonomy of Affect."

9. Mazzarella, "Affect"; Larkin, "Politics and Poetics"; Harvey and Knox, "Enchantments of Infrastructure."

10. Anand, Gupta, and Appel, *Promise of Infrastructure*.

11. Matlon, "This Is How We Roll"; Weiss, *Street Dreams*; Newell, *Modernity Bluff*.

12. Article 39 of the Constitution of the Republic of Uganda reads, in full, "Every Ugandan has a right to a clean and healthy environment." Chapter 3 ("Citizenship"), Article 17 ("Duties of a Citizen"), states that "It is the duty of every citizen of Uganda . . . (j) to create and protect a clean and healthy environment" Republic of Uganda, "Constitution of the Republic of Uganda."

13. Doherty, "Motorcycle Taxis"; S. T. Brown, "Kampala's Sanitary Regime"; McFarlane and Silver, "Navigating the City."

CHAPTER 12. DEVELOPMENTAL RESPECTABILITY

1. Asad, *Anthropology and the Colonial Encounter*, 5.

2. Pierre, *Predicament of Blackness*, 89–90.

3. West and Sanders, *Transparency and Conspiracy*.

4. These names are just the most prominent cases of anti-Black police violence in the United States. Writing in 2019, on the fifth anniversary of the rise of the Black Lives Matter movement, Keeanga-Yamahtta Taylor pointed to the breadth of this violence: "In the five years since Mike Brown Jr. was murdered and the streets of Ferguson, Missouri erupted, police across the United States have killed more than four thousand people, a quarter of them African American." K. Taylor, "Five Years Later."

5. Morris, *Close Kin and Distant Relatives*, 5.

6. Gray, "Subject to Respectability"; Cooper, *Beyond Respectability*.

7. Gaines, *Uplifting the Race*.

8. Gaines, xvii.

9. White, *Dark Continent of Our Bodies*.

10. Rhodes, "Pedagogies of Respectability."

11. Wolcott, *Remaking Respectability*.

12. Morris, *Close Kin and Distant Relatives*, 29.

13. Higginbotham, *Righteous Discontent*.

14. Gaines, *Uplifting the Race*, 158.

15. Díaz, "Racial Trust"; Kelley, "Not What We Seem."

16. K. Ferguson, *Black Politics*, 192.

17. Solomon, "'Ghetto Is a Gold Mine,'" 76.

18. Lee and Hicken, "Death by a Thousand Cuts."

19. M. Alexander, *New Jim Crow*, 212–15.

20. Morris, *Close Kin and Distant Relatives*, 8.

21. *New York Times*, "Michael Brown."

22. White, *Dark Continent of Our Bodies*, 123.

23. Zimmerman, *Alabama in Africa*.

24. Comaroff and Comaroff, *Ethnography*, 70.

25. Hansen, *African Encounters with Domesticity*; Comaroff and Comaroff, *Revelation and Revolution*.

26. Burke, *Lifebuoy Men, Lux Women*; Swanson, "Sanitation Syndrome."

27. Vaughan, *Curing Their Ills*; McClintock, *Imperial Leather*; Newell, *Histories of Dirt*.

28. West, *Rise of an African Middle Class*.

29. N. Musisi, "Colonial and Missionary Education," 186; Tiberondwa, *Missionary Teachers*.

30. N. Musisi, "Colonial and Missionary Education."

31. Kyomuhendo and McIntosh, *Women, Work and Domestic Virtue*, 56.

32. Summers, "Intimate Colonialism."

33. Wallman, *Kampala Women Getting By*; Parikh, "From Auntie to Disco"; Whyte, *Second Chances*; Cruz, "In Search of Safety."

34. White, *Dark Continent of Our Bodies*, 35.

35. Parkin, *Neighbors and Nationals*; Southall and Gutkind, *Townsmen in the Making*.

36. Ogden, "'Producing' Respect."

37. Ogden, 178.

38. Obbo, *African Women*.

39. Kyomuhendo and McIntosh, *Women, Work and Domestic Virtue*.

40. Summers, "Radical Rudeness"; Diabate, *Naked Agency*.

41. Nyanzi, *No Roses from My Mouth*.

42. Gutkind, *Royal Capital of Buganda*.

43. Earle, *Colonial Buganda*; Karlström, "Imagining Democracy."

44. Thanks to Sandra Calkins and Tyler Zoanni for their insight on this point and for bringing Mukasa's text to my attention.

45. Mukasa, Rowe, and Kabazza, *Simuda Nyuma*, 1.

46. Karlström, "Modernity and Its Aspirants"; Mehta, *Liberalism and Empire*.

47. Pierre, *Predicament of Blackness*, 74.

48. Pierre, 75.

49. Berlant, *Cruel Optimism*.

CHAPTER 13. WASTE IN TIME

1. The government bought these jets in 2011 from Russia at a cost of US$740 million, raising questions about national priorities in the national budget. *East African*, "$740m Fighter Jets Scam Sneaks under the Radar."

2. Human Rights Watch, *Letting the Big Fish Swim*; *New Vision*, "OPM"; *New Vision*, "Germany Halts Aid."

3. *New Vision*, "Museveni Explains Paraffin"; *New Vision*, "Kerosene Tax"; *New Vision*, "Budget: Experts React."

4. Bichachi, "Village Still Lives."

5. Golooba-Mutebi, "Our Post-Colonial 'Transformation.'"

6. Golooba-Mutebi.

7. A. Alexander, "Disciplining Method."

8. Bichachi, "Village Still Lives."

9. Bichachi.

10. Rosaldo, *Culture and Truth*, 68; see also Dlamini, *Native Nostalgia*.

11. Omolo-Okalebo et al., "Planning Ideas"; Gutkind, *Royal Capital of Buganda*; Myers, *Verandahs of Power*.

12. Naggaga, "Were African Countries Ready."

13. Mamdani, *Politics and Class Formation*.

14. E. Taylor, "Claiming Kabale."

15. Hundle, "African Asians and South Asians."

16. Pierre, *Predicament of Blackness*, 35.

17. Finnstrom, *Living with Bad Surroundings*, 79–81; Amone, "Creation of Acholi Military Ethnocracy."

18. Mamdani, *Imperialism and Fascism*, 10.

19. Laruni, "From the Village to Entebbe," 25–27; Farelius, "Where Does the Hamite Belong?," 110; Pierre, "Slavery, Anthropological Knowledge."

20. Pier, "Branded Arena."

21. Brisset-Foucault, "Citizenship of Distinction."

22. Foucault, *Society Must Be Defended*; Chari, "State Racism and Biopolitical Struggle."

23. Mbembe, "Necropolitics"; Singh, *Black Is a Country*; Stoler, *Race and the Education of Desire*; Weheliye, *Habeas Viscus*.

24. Foucault, *Society Must Be Defended*, 255.

25. Singh, *Black Is a Country*, 223.

26. Gidwani, *Capital, Interrupted*, 11.

27. Smith, *Bewitching Development*, 6; See also Mehta, *Liberalism and Empire*.

28. Churchill, *Kill the Indian, Save the Man*; Thiong'o, *Decolonising the Mind*.

29. Fanon, *Wretched of the Earth*, 129.

30. Fanon, 131.

31. Fanon, 127.

32. Fanon, 156.

33. Kamya, "Ugandans Are Poor."

34. Museveni, "Fanon's Theory of Violence"; *New Vision*, "Sins of Uganda"; *New Vision*, "Lazy Workforce"; *New Vision*, "Tired of Cheap Politicking"; *New Vision*, "Lazy, Thieving Ugandans."

35. Museveni, *Sowing the Mustard Seed*, 35.

36. World Bank, "World Bank Open Data."

37. Brennan, *Taifa*, 118–19.

38. Museveni, "The old man with the hat."

39. Elkan, *Migrants and Proletarians*.

40. Mwenda, "How to Change Kampala"; Kamya, "Ugandans Are Poor."

CHAPTER 14. CLEAN HEARTS, DIRTY HANDS

1. Kampala City Yange, "About Us."

2. Quotes are from my interviews, including with two leaders of Kampala City Yange, pseudonymized as Godfrey Katete and Ezra Kalungi as well as from conversations with other KCY staff during cleaning exercises.

3. Os Suna, "Kampala City Yange," BuzyBee Ent / Pro Kayz, Georgeous Films Inc., directed by Nolton/Noltomax, 2013, https://www.youtube.com/watch?v=enZt2mYjmHo.

4. Kony 2012 was an internet-based campaign organized by the NGO Invisible Children to build an international movement to raise awareness—through viral videos and social marketing—about Joseph Kony and the Lord's Resistance Army in Northern Uganda in order to (through mechanisms left unstated) lead to his capture and trial. The campaign was widely critiqued for being infantilizing and radically reductive—reducing the conflict to the evil of a single man—and for calling for further militarization of the region without taking into account the role of Ugandan security forces in human rights abuses and land grabbing. Branch, "Dangerous Ignorance"; Izama, "Kony Is Not the Problem"; T. Cole, "White Savior Industrial Complex," 340.

5. Fassin, "Humanitarianism as a Politics of Life," 500.

6. Malkki, *Need to Help*; Dahl, "Too Fat"; Bornstein, *Spirit of Development*; Heron, *Desire for Development*.

7. Roy and Crane, *Territories of Poverty*.

8. *The Onion*, "6-Day Visit."

9. Kampala City Yange, "Awards."

10. Kampala City Yange.

11. Stallybrass and White, *Politics and Poetics*.

12. *In Kampala*, "A Clean Sweep."

13. *New Vision,* "Kampala a Mountain of Garbage."

14. Luke 22:27.

15. Nation, "Footwashing."

16. *New Vision,* "We Are All Responsible."

17. *New Vision.*

18. Adengo, "Love of a Clean City."

19. These accounts of urban poverty were also framed by racial difference. The "other half" in the tenements of New York documented in Jacob Riis's *How the Other Half Lives,* for instance, was comprised of European immigrants still in the process of becoming white. Equally, in the industrial slums of Britain, the most abject conditions described in Engels's *The Condition of the Working Class in England* are found among Irish immigrants racialized as other to such an extent that the socialist reformer disdained Irish workers as uncivilized. Racialized others, split along Victorian mores of respectability, constituted the lumpenproletariat that Marx and Engels dismissed in order to cleanse their notion of the pure English proletariat as the revolutionary subject. These ruptures posited within the population, between the working class and the lumpen, evidenced by everyday degeneracy and paired with differentiated assumptions about historical agency, epitomize Foucault's notion of state racism. See Bussard, "Dangerous Classes"; Ignatiev, *How the Irish Became White*; Roediger, *Working toward Whiteness*; Foucault, *Society Must Be Defended*; Stallybrass, "Marx and Heterogeneity."

20. Adengo, "Love of a Clean City."

21. Adengo.

22. Adengo.

23. Adengo.

24. *New Vision,* "Pictorial."

25. Mwenda, "How to Change Kampala."

26. *New Vision.* "Pictorial."

27. Brisset-Foucault, "Citizenship of Distinction."

28. Doherty, "Motorcycle Taxis"; Scherz, *Having People*; Zoanni, "Appearances of Disability."

29. *New Vision,* "Kampala on Its Way."

30. *The Observer,* "City Dwellers Get Brooms."

31. Mwenda, "How to Change Kampala."

32. Locke, *Two Treatises of Government*; Macpherson, *Political Theory of Possessive Individualism.*

33. Kampala City Yange, "About Us."

34. Ghertner, "Nuisance Talk."

35. Ghertner, *Rule by Aesthetics*; Ticktin, *Casualties of Care.*

36. Fredericks, *Garbage Citizenship,* 60–74; Diouf, "Urban Youth and Senegalese Politics."

37. Elshahed, "Tahrir Square."

38. Redfield and Robins, "Index of Waste."

39. T. Cole, "White Savior Industrial Complex," 340.

CONCLUSION

1. Gunston, "Planning and Construction."

2. Whitehouse, "Kenya and Uganda Railway."

3. Obita, "Justice in Land," 24.

4. Mutambatsere et al., "Role for Multilateral Development Banks."

5. *Daily Monitor*, "Harrowing Tales."

6. *Chimp Reports*, "Teargas as Railway Reserve Squatters Protest."

7. *The Independent*, "Remembering the KCCA Evictions."

8. *The East African*, "SGR Future in Doubt."

9. Patel and Moore, *History of the World*.

10. Magubane, *Construction of the Dispensable Other*; Mbembe, "Necropolitics"; Hage, *Alter-Politics*.

11. Bullard, *Dumping in Dixie*; Chari, "Detritus in Durban"; Liboiron, "Waste Colonialism."

12. Cruvinel et al., "Health Conditions and Occupational Risks"; Samarth, *Occupational Health of Waste Pickers*.

13. Solomon, "Ghetto Is a Gold Mine"; Chatterjee, "Labors of Failure."

14. Voyles, *Wastelanding*.

15. Ghertner, *Rule by Aesthetics*.

Bibliography

Abrahamsen, Rita, and Gerald Bareebe. "Uganda's 2016 Elections: Not Even Faking It Anymore." *African Affairs* 115, no. 461 (2016): 751–65.

Adengo, Jonathan. "For the Love of a Clean City." *Daily Monitor*. October 7, 2012.

Adichie, Chimamanda Ngozi. "An Interview with *'Americanah'* Author Chimamanda Ngozi Adichie." By Michele Norris. Aspen Institute, 2014. https://archive.org/details/An_Interview_with_Americanah_Author_Chimamanda_Ngozi_Adichie.

Agrawal, Arun. *Environmentality: Technologies of Government and the Making of Subjects.* Durham, NC: Duke University Press, 2005.

Ahmed, Sara. *Queer Phenomenology: Objects, Orientations, Others.* Durham, NC: Duke University Press, 2006.

Akaki, Tony. *Mabira Forest Giveaway: A Path to Degenerative Development.* iUniverse, 2011.

Alaimo, Stacy. *Exposed: Environmental Politics and Pleasures in Posthuman Times.* Minneapolis: University of Minnesota Press, 2016.

Alexander, Amanda. "A Disciplining Method for Holding Standards Down: How the World Bank Planned Africa's Slums." *Review of African Political Economy* 39, no. 134 (2012): 590–613.

Alexander, Catherine, and Joshua Reno, eds. *Economies of Recycling: The Global Transformation of Materials, Values and Social Relations.* London: Zed Books, 2012.

Alexander, Michelle. *The New Jim Crow: Mass Incarceration in the Age of Colorblindness*. New York: New Press, 2010.

Althusser, Louis. *On the Reproduction of Capitalism: Ideology and Ideological State Apparatuses*. New York: Verso, 2014.

Alvarado-Esquivel, Cosme. "*Helicobacter Pylori* Infection in Waste Pickers: A Case Control Seroprevalence Study." *Gastroenterology Research* 6, no. 5 (2013): 174–79.

Amnesty International. "Stifling Dissent: Restrictions on the Rights to Freedom of Expression and Peaceful Assembly in Uganda." 2011. https://www .amnesty.org/en/documents/AFR59/016/2011/en.

Amone, Charles. "The Creation of Acholi Military Ethnocracy in Uganda, 1862 to 1962." *International Journal of Liberal Arts and Social Science* 2, no. 3 (2014): 141–50.

Anand, Nikhil. *Hydraulic City: Water and the Infrastructures of Citizenship in Mumbai*. Durham, NC: Duke University Press, 2017.

Anand, Nikhil, Akhil Gupta, and Hannah Appel, eds. *The Promise of Infrastructure*. Durham, NC: Duke University Press, 2018.

Appelblad Fredby, Jenny, and David Nillson. "From 'All for Some' to 'Some for All'? A Historical Geography of Pro-Poor Water Provision in Kampala." *Journal of East African Studies* 7, no. 1 (2012): 40–57.

Apter, Andrew. *The Pan-African Nation: Oil and the Spectacle of Culture in Nigeria*. Chicago: University of Chicago Press, 2005.

Apter, David. *The Political Kingdom in Uganda: A Study in Bureaucratic Nationalism*. Princeton, NJ: Princeton University Press, 1961.

Argyrou, Vassos. "'Keep Cyprus Clean': Littering, Pollution, and Otherness." *Cultural Anthropology* 12, no. 2 (1997): 159–78.

Asad, Talal, ed. *Anthropology and the Colonial Encounter*. Atlantic Highlands, NJ: Humanities Press, 1973.

Baral, Anna. "'Like the Chicken and the Egg': Market Vendors and the Dilemmas of Neoliberal Urban Planning in Re-Centralised Kampala (Uganda)." *Kritsk Etnografi* 2, no. 1–2 (2019): 16.

Barnard, Alex. *Freegans: Diving into the Wealth of Food Waste in America*. Minneapolis: University of Minnesota Press, 2016.

Barr, Abigail, Marcel Fafchamps, and Trudy Owens. "The Governance of Non-Governmental Organizations in Uganda." *World Development* 33, no. 4 (2005): 657–79.

Barry, Andrew, Thomas Osborne, and Nikolas Rose, eds. *Foucault and Political Reason: Liberalism, Neo-Liberalism and Rationalities of Government*. Chicago: University of Chicago Press, 1996.

Bashford, Allison. *Imperial Hygiene: A Critical History of Colonialism, Nationalism and Public Health*. New York: Palgrave Macmillan, 2005.

Baudouin, Axel, Camilla Bjerkli, Yigalem Habtemariam, and Zelalem Fenta Chekole. "Between Neglect and Control: Questioning Partnerships and the Integration of Informal Actors in Public Solid Waste Management in Addis Ababa, Ethiopia." *African Studies Quarterly* 11, no. 2–3 (2010): 29–42.

Bauman, Zygmunt. *Wasted Lives: Modernity and Its Outcasts.* Malden, MA: Polity, 2004.

Baviskar, Amita, and Raka Ray, eds. *Elite and Everyman: The Cultural Politics of the Indian Middle Classes.* New York: Routledge, 2011.

Bayat, Asef. *Life as Politics: How Ordinary People Change the Middle East.* Stanford, CA: Stanford University Press, 2010.

BBC News. "Protesters Fish in Uganda Pothole." June 8, 2010, sec. Africa. https://www.bbc.com/news/10268958.

Benezra, Amber, Joseph DeStefano, and Jeffrey I. Gordon. "Anthropology of Microbes." *Proceedings of the National Academy of Sciences* 109, no. 17 (2012): 6378–81.

Benjamin, Solomon. "Occupancy Urbanism: Radicalizing Politics and Economy beyond Policy and Programs." *International Journal of Urban and Regional Research* 32, no. 3 (2008): 719–29.

Bennett, Jane. *Vibrant Matter: A Political Ecology of Things.* Durham, NC: Duke University Press, 2009.

Berlant, Lauren. *Cruel Optimism.* Durham, NC: Duke University Press, 2011.

Bernal, Victoria, and Inderpal Grewal, eds. *Theorizing NGOs: States, Feminisms, and Neoliberalism.* Durham, NC: Duke University Press, 2014.

Bhan, Gautam. "A Bold Plan to House 100 Million People." TED Talks India, December 11, 2017. https://www.ted.com/talks/gautam_bhan_a_bold_plan _to_house_100_million_people?utm_campaign=tedspread&utm_medium= referral&utm_source=tedcomshare.

Bichachi, Charles Odoobo. "The Village Still Lives with Us in the City." *The Observer,* Uganda, October 18, 2012.

Bize, Amiel. "The Right to the Remainder: Gleaning in the Fuel Economies of East Africa's Northern Corridor." *Cultural Anthropology* 35, no. 3 (2020): 462–86.

Bjerkli, Camilla. "Governance on the Ground: A Study of Solid Waste Management in Addis Ababa, Ethiopia." *International Journal of Urban and Regional Research* 37, no. 4 (2013): 1273–87.

Bloom, Peter, Stephan Miescher, and Takiywaa Manuh, eds. *Modernization as Spectacle in Africa.* Bloomington: Indiana University Press, 2014.

Boeck, Filip De. "Inhabiting Ocular Ground: Kinshasa's Future in the Light of Congo's Spectral Urban Politics." *Cultural Anthropology* 26, no. 2 (2011): 263–86.

———. "Spectral Kinshasa: Building the City through an Architecture of Words."
In *Urban Theory Beyond the West: A World of Cities*, edited by Tim Edensor
and Mark Jayne, 311–28. New York: Routledge, 2012.

Boeck, Filip De, and Sammy Baloji. *Suturing the City: Living Together in
Congo's Urban Worlds*. London: Autograph, 2016.

Bornstein, Erica. *The Spirit of Development: Protestant NGOs, Morality, and
Economics in Zimbabwe*. Stanford, CA: Stanford University Press, 2003.

Bornstein, Erica, and Aradhana Sharma. "The Righteous and the Rightful: The
Technomoral Politics of NGOs, Social Movements, and the State in India."
American Ethnologist 43, no. 1 (2016): 76–90.

Branch, Adam. "Dangerous Ignorance: The Hysteria of Kony 2012." *Al Jazeera*,
March 12, 2012. https://www.aljazeera.com/opinions/2012/3/12/dangerous
-ignorance-the-hysteria-of-kony-2012.

Branch, Adam, and Zachariah Mampilly. *Africa Uprising: Popular Protest and
Political Change*. London: Zed Books, 2015.

Brautigam, Deborah, and Tang Ziaoyang. "African Shenzhen: China's Special
Economic Zones in Africa." *Journal of Modern African Studies* 49, no. 1
(2011): 27–54.

Brennan, James. *Taifa: Making Nation and Race in Urban Tanzania*. Athens:
Ohio University Press, 2012.

Brisset-Foucault, Florence. "A Citizenship of Distinction in the Open Radio
Debates of Kampala." *Africa* 83, no. 2 (2013): 227–50.

Brown, Stephanie Terreni. "Kampala's Sanitary Regime: Whose Toilet Is It
Anyway?" In *Infrastructural Lives: Urban Infrastructure in Context*, edited
by Stephen Graham and Colin McFarlane, 154–73. New York: Routledge,
2015.

———. "Planning Kampala: Histories of Sanitary Intervention and In/Formal
Spaces." *Critical African Studies* 6, no. 1 (2014): 71–90.

Brown, Stephen. "Michel Serres: Science, Translation and the Logic of the
Parasite." *Theory, Culture & Society* 19, no. 3 (2002): 1–27.

Brunner, Franziska S., Jaime M. Anaya-Rojas, Blake Matthews, and Christophe
Eizaguirre. "Experimental Evidence That Parasites Drive Eco-Evolutionary
Feedbacks." *Proceedings of the National Academy of Sciences* 114, no. 14
(2017): 3678–83.

Bullard, Robert. *Dumping in Dixie: Race, Class, and Environmental Quality*.
Boulder, CO: Westview Press, 2000.

Burke, Timothy. *Lifebuoy Men, Lux Women: Commodification, Consumption,
and Cleanliness in Modern Zimbabwe*. Durham, NC: Duke University
Press, 1996.

Burton, Andrew. *African Underclass: Urbanisation, Crime and Colonial Order
in Dar es Salaam*. Athens: Ohio University Press, 2005.

Bussard, Robert. "The 'Dangerous Classes' of Marx and Engels: The Rise of the Idea of the Lumpenproletariat." *History of European Ideas* 8, no. 6 (1987): 675–92.

Carbone, Giovanni. *No Party Democracy? Ugandan Politics in Comparative Perspective*. Boulder, CO: Lynne Rienner, 2008.

Chalfin, Brenda. "Waste Work and the Dialectics of Precarity in Urban Ghana: Durable Bodies and Disposable Things." *Africa* 89, no. 3 (2019): 499–520.

Chari, Sharad. "Detritus in Durban: Polluted Environs and the Biopolitics of Refusal." In *Imperial Debris: On Ruins and Ruination*, edited by Ann Laura Stoler, 131–61. Durham, NC: Duke University Press, 2013.

———. "State Racism and Biopolitical Struggle: The Evasive Commons in Twentieth-Century Durban, South Africa." *Radical History Review* 2010, no. 108 (2010): 73–90.

Chatterjee, Syantani. "The Labors of Failure: Labor, Toxicity, and Belonging in Mumbai." *International Labor and Working-Class History* 95 (2019): 49–75.

Chattopadhyay, Swati. *Unlearning the City: Infrastructures in a New Optical Field*. Minneapolis: University of Minnesota Press, 2012.

Checker, Melissa. *Polluted Promises: Environmental Racism and the Search for Justice in a Southern Town*. New York: New York University Press, 2005.

Chikarmane, Poornima. *Integrating Waste Pickers into Municipal Solid Waste Management in Pune, India*. WIEGO Policy Brief (Urban Policies) no. 8. Manchester, UK: Women in Informal Employment: Globalizing and Organizing, 2012. www.wiego.org/sites/default/files/publications/files /Chikarmane_WIEGO_PB8.pdf.

Chimp Reports. "Teargas as Railway Reserve Squatters Protest." August 4, 2014. https://chimpreports.com/teargas-as-railway-reserve-occupants-protest.

Chu, Julie. "When Infrastructures Attack: The Workings of Disrepair in China." *American Ethnologist* 41, no. 2 (2014): 351–67.

Churchill, Ward. *Kill the Indian, Save the Man: The Genocidal Impact of American Indian Residential Schools*. San Francisco: City Lights Publishers, 2004.

Cointreau, Sandra. "Occupational and Environmental Health Issues of Solid Waste Management: Special Emphasis on Middle- and Lower-Income Countries." Urban Papers. Washington, DC: World Bank, 2006. http:// documents.worldbank.org/curated/en/679351468143072645/Occupational -and-environmental-health-issues-of-solid-waste-management-special -emphasis-on-middle-and-lower-income-countries.

———. "Sanitary Landfill Design and Siting Criteria." World Bank Urban Infrastructure Note. Washington, DC: World Bank, 1996. http://documents1 .worldbank.org/curated/en/461871468139209227/pdf /337800revoLandfillsitingdesign.pdf.

Cole, Georgia. "Historical Backlash or Contemporary Realities for Uganda's Asian Population?" *African Arguments* (blog), November 11, 2013. https://africanarguments.org/2013/11/historical-backlash-or-contemporary-realities-for-ugandas-asian-populations-by-georgia-cole.

Cole, Teju. "The White Savior Industrial Complex." In *Known and Strange Things*, 340–49. New York: Random House, 2016.

Comaroff, John L., and Jean Comaroff. *Ethnography and the Historical Imagination*. Boulder, CO: Westview Press, 1992.

———. *Of Revelation and Revolution: The Dialectics of Modernity on a South African Frontier*, Vol. 2. Chicago: University of Chicago Press, 1997.

Cooper, Brittney. *Beyond Respectability: The Intellectual Thought of Race Women*. Urbana, IL: University of Illinois Press, 2017.

Cooper, Frederick. *Africa in the World: Capitalism, Empire, Nation-State*. Cambridge, MA: Harvard University Press, 2014.

Cram, Shannon. "Wild and Scenic Wasteland: Conservation Politics in the Nuclear Wilderness." *Environmental Humanities* 7, no. 1 (2016): 89–105.

Crewe, Emma, and Priyanthi Fernando. "The Elephant in the Room: Racism in Representations, Relationships and Rituals." *Progress in Development Studies* 6, no. 1 (2006): 40–54.

Cruvinel, Vanessa Resende Nogueira, Carla Pintas Marques, Vanessa Cardoso, Maria Rita Carvalho Garbi Novaes, Wildo Navegantes Araújo, Antonia Angulo-Tuesta, Patrícia Maria Fonseca Escalda, Dayani Galato, Petruza Brito, and Everton Nunes da Silva. "Health Conditions and Occupational Risks in a Novel Group: Waste Pickers in the Largest Open Garbage Dump in Latin America." *BMC Public Health* 19, no. 1 (2019): 581.

Cruz, Serena. "In Search of Safety, Negotiating Everyday Forms of Risk: Sex Work, Criminalization, and HIV/AIDS in the Slums of Kampala." PhD diss., Florida International University, 2015.

Dabiri, Emma. "Why I'm Not an Afropolitan." *Africa Is a Country* (blog), January 21, 2014. https://africasacountry.com/2014/01/why-im-not-an-afropolitan.

Dahl, Bianca. "Too Fat to Be an Orphan: The Moral Semiotics of Food Aid in Botswana." *Cultural Anthropology* 29, no. 4 (2014): 626–47.

Daily Monitor. "Besigye—I Will Continue Walking to Work." May 14, 2011.

———. "Ex-PS Explains Naguru City Project Delay." October 18, 2013.

———. "Fascism, City, Country, KCCA." February 5, 2012.

———. "Government Tackles Flooding in Zzana Town." November 21, 2007.

———. "Harrowing Tales of Living in a Railway Reserve." March 22, 2019.

———. "IGG Denies Blocking Naguru Estate Work." July 2, 2014.

———. "Jennifer Musisi, Loathed and Admired in the Same Measure." December 4, 2011.

———. "KCCA Reveals Shs485b Key Development Plans for Kampala." December 24, 2014.

———. "KCCA Tests Passenger Train." February 5, 2015.

———. "Museveni Commissions 1,700 New Housing Units in Naguru." October 15, 2013.

———. "Musisi Ordered to Open KCCA Offices." November 30, 2013.

———. "Nakawa-Naguru Families Stranded as Evictions Begin." July 5, 2011.

———. "New KCCA Law Shouldn't Take Away Voters Rights." November 13, 2015.

———. "New Taxi Park Renovation: How the People Are Coping." October 4, 2012.

———. "Railway Eviction Case Adjourned." October 14, 2014.

———. "Speaker Kadaga Promises to Revive Shelved Gay Bill." October 31, 2012.

Davis, Heather. "Toxic Progeny: The Plastisphere and Other Queer Futures." *PhiloSOPHIA* 5, no. 2 (2015): 231–50.

Davis, Mike. *Planet of Slums*. New York: Verso, 2006.

Decker, Alicia. *In Idi Amin's Shadow: Women, Gender, and Militarism in Uganda*. Athens: Ohio University Press, 2014.

De León, Jason. *The Land of Open Graves: Living and Dying on the Migrant Trail*. Oakland: University of California Press, 2015.

Demaria, Federico, and Seth Schindler. "Contesting Urban Metabolism: Struggles over Waste-to-Energy in Delhi, India." *Antipode* 48, no. 2 (2016): 293–313.

Desai, Ashwin. *We Are the Poors: Community Struggles in Post-Apartheid South Africa*. New York: New York University Press, 2002.

Diabate, Naminata. *Naked Agency: Genital Cursing and Biopolitics in Africa*. Durham, NC: Duke University Press, 2020.

Díaz, Sara. "'A Racial Trust': Individualist, Eugenicist, and Capitalist Respectability in the Life of Roger Arliner Young." *Souls* 18, no. 2–4 (2016): 235–62.

Dill, Brian. *Fixing the African State: Recognition, Politics, and Community-Based Development in Tanzania*. New York: Palgrave MacMillan, 2013.

Diouf, Mamadou. "Urban Youth and Senegalese Politics: Dakar 1988–1994." *Public Culture* 8, no. 2 (1996): 225–49.

Dlamini, Jacob. *Native Nostalgia*. Auckland Park, South Africa: Jacana Media, 2009.

Doherty, Jacob. "Capitalizing Community: Waste, Wealth, and (Im)Material Labor in a Kampala Slum." *International Labor and Working-Class History* 95 (2019): 95–113.

———. "Filthy Flourishing: Para-Sites, Animal Infrastructure, and the Waste Frontier in Kampala." *Current Anthropology* 60, no. S20 (2019): S321–32.

———. "Life (and Limb) in the Fast-Lane: Disposable People as Infrastructure in Kampala's Boda Boda Industry." *Critical African Studies* 9, no. 1 (2017): 192–209.

———. "Maintenance Space: The Political Authority of Garbage in Kampala, Uganda." *Current Anthropology* 60, no. 1 (2019): 24–46.

———. "Motorcycle Taxis, Personhood, and the Moral Landscape of Mobility." *Geoforum*, 2020.

———. "Why Is This Trash Can Yelling at Me? Big Bellies and Clean Green Gentrification." *Anthropology Now* 10, no. 1 (2018): 93–101.

Doherty, Jacob, and Kate Brown. "Labor Laid Waste: An Introduction to the Special Issue on Waste Work." *International Labor and Working-Class History* 95 (2019): 1–17.

Donovan, Kevin. "Colonizing the Future." *Boston Review*, September 23, 2020. http://bostonreview.net/class-inequality-global-justice/kevin-p-donovan -colonizing-future.

Doron, Assa, and Robin Jeffrey. *Waste of a Nation: Garbage and Growth in India*. Cambridge, MA: Harvard University Press, 2018.

Douglas, Mary. *Purity and Danger: An Analysis of the Concepts of Pollution and Taboo*. New York: Routledge, 1966.

Earle, Jonathon. *Colonial Buganda and the End of Empire: Political Thought and Historical Imagination in Africa*. Cambridge, UK: Cambridge University Press, 2017.

The East African. "$740m Fighter Jets Scam Sneaks under the Radar." April 4, 2011.

———. "SGR Future in Doubt as Governments Revert to Old Metre Gauge Railways." *The East African*. June 19, 2019.

Elkan, Walter. *Migrants and Proletarians: Urban Labour in the Economic Development of Uganda*. New York: Oxford University Press, 1960.

Elshahed, Mohamed. "Tahrir Square: Social Media, Public Space." *Places Journal*, 2011. https://placesjournal.org/article/tahrir-square-social-media -public-space.

Escobar, Arturo. *Encountering Development: The Making and Unmaking of the Third World*. Princeton, NJ: Princeton University Press, 1995.

Evans, Brad, ed. "Disposable Life." Histories of Violence. https://www.histories ofviolence.com/disposablelife.

Evans, Brad, and Henry Giroux. *Disposable Futures: The Seduction of Violence in the Age of Spectacle*. San Francisco: City Lights Publishers, 2015.

Fahmi, Wael Salah. "The Impact of Privatization of Solid Waste Management on the Zabaleen Garbage Collectors of Cairo." *Environment and Urbanization* 17, no. 2 (2005): 155–70.

Fanon, Frantz. *The Wretched of the Earth*. New York: Penguin, 1967.

Farelius, Birgitta. "Where Does the Hamite Belong?" *Nomadic Peoples*, no. 32 (1993): 107–18.

Fassin, Didier. "Humanitarianism as a Politics of Life." *Public Culture* 19, no. 3 (2007): 499–520.

Ferguson, James. *The Anti-politics Machine: "Development," Depoliticization, and Bureaucratic Power in Lesotho.* Minneapolis: University of Minnesota Press, 1994.

———. *Give a Man a Fish: Reflections on the New Politics of Distribution.* Durham, NC: Duke University Press, 2015.

———. *Global Shadows: Africa in the Neoliberal World Order.* Durham, NC: Duke University Press, 2006.

Ferguson, James, and Akhil Gupta. "Spatializing States: Toward an Ethnography of Neoliberal Governmentality." *American Ethnologist* 29, no. 4 (2002): 981–1002.

Ferguson, Karen. *Black Politics in New Deal Atlanta.* Chapel Hill: University of North Carolina Press, 2002.

Fergutz, Oscar, et al. "Developing Urban Waste Management in Brazil with Waste Picker Organizations." *Environment and Urbanization* 23, no. 2 (2011): 597–608.

Finnstrom, Sverker. *Living with Bad Surroundings: War, History, and Everyday Moments in Northern Uganda.* Durham, NC: Duke University Press, 2008.

Foucault, Michel. *"Society Must Be Defended": Lectures at the College de France, 1975–6.* New York: Picador, 2003.

Frearson, Amy. "David Adjaye Designs Office Campus for New 65-Hectare Development in Uganda." *Dezeen* (blog), January 2, 2014. https://www.dezeen.com/2014/01/02/david-adjaye-office-campus-kampala-uganda.

Fredericks, Rosalind. *Garbage Citizenship: Vital Infrastructures of Labor in Dakar, Senegal.* Durham, NC: Duke University Press, 2018.

Gabrys, Jennifer. "Sink: The Dirt of Systems." *Environment and Planning D* 27 (2009): 666–81.

Gabrys, Jennifer, Gay Hawkins, and Mike Michael, eds. *Accumulation: The Material Politics of Plastic.* Abingdon, UK: Routledge, 2013.

Gaines, Kevin. *Uplifting the Race: Black Leadership, Politics, and Culture in the Twentieth Century.* Chapel Hill: University of North Carolina Press, 1996.

Gandy, Matthew. "Learning from Lagos." *New Left Review* 33, no. 37–53 (2005).

German Technical Cooperation Agency. "Solid Waste Disposal, Kampala," March 1990.

Ghertner, D. Asher. "Nuisance Talk and the Propriety of Property: Middle Class Discourses of a Slum-Free Delhi." *Antipode* 44, no. 4 (2012): 1161–87.

———. *Rule by Aesthetics: World-Class City Making in Delhi.* New York: Oxford University Press, 2015.

Gidwani, Vinay. *Capital, Interrupted: Agrarian Development and the Politics of Work in India.* Minneapolis: University of Minnesota Press, 2008.

———. "The Work of Waste: Inside India's Infra-Economy." *Transactions of the Institute of British Geographers* 40, no. 4 (2015): 575–95.

Giles, David Boarder. "The Anatomy of a Dumpster: Abject Capital and the Looking Glass of Value." *Social Text* 32, no. 1 (2014): 93–113.

Gill, Kaveri. *Of Poverty and Plastic: Scavenging and Scrap Trading in India's Urban Informal Economy*. New York: Oxford University Press, 2010.

Gilmore, Ruth Wilson. *Golden Gulag: Prisons, Surplus, Crisis, and Opposition in Globalizing California*. Berkeley: University of California Press, 2007.

Goldstein, Alyosha. *Poverty in Common: The Politics of Community Action during the American Century*. Durham, NC: Duke University Press, 2012.

Goldstein, Joshua. "Waste." In *The Oxford Handbook of the History of Consumption*, edited by Frank Trentmann, 326–47. Oxford: Oxford University Press, 2012.

Golooba-Mutebi, Frederick. "Devolution and Outsourcing of Municipal Services in Kampala City, Uganda: An Early Assessment." *Public Administration and Development* 23, no. 5 (2003): 405–18.

———. "Our Post-Colonial 'Transformation': Private Cleanliness Amid Public Filth." *The East African*. October 2, 2011.

Gombay, Christine. "Eating Cities: Urban Management and Markets in Kampala." *Cities* 11, no. 2 (1994): 86–94.

Goodfellow, Tom. "'The Bastard Child of Nobody'? Anti-Planning and the Institutional Crisis in Contemporary Kampala." Working Papers Series no. 2. Working paper no. 67. London: Crisis States Research Centre, 2010. https://ethz.ch/content/specialinterest/gess/cis/center-for-securities-studies/en/services/digital-library/publications/publication.html/112104.

———. "The Institutionalisation of 'Noise' and 'Silence' in Urban Politics: Riots and Compliance in Uganda and Rwanda." *Oxford Development Studies* 41, no. 4 (2013): 436–54.

———. "Legal Manoeuvres and Violence: Law Making, Protest and Semi-Authoritarianism in Uganda." *Development and Change* 45, no. 4 (2014): 753–76.

Gordillo, Gastón. *Rubble: The Afterlife of Destruction*. Durham, NC: Duke University Press, 2014.

Gore, Christopher, and Nansozi Muwanga. "Decentralization Is Dead, Long Live Decentralization! Capital City Reform and Political Rights in Kampala, Uganda." *International Journal of Urban and Regional Research* 38, no. 6 (2014): 2201–16.

Graber, Lauren, Daniel Asher, Natasha Anandaraja, Richard Bopp, Karen Merrill, Mark Cullen, Samuel Luboga, and Leonardo Trasande. "Childhood Lead Exposure after the Phaseout of Leaded Gasoline: An Ecological Study of School-Age Children in Kampala, Uganda." *Environmental Health Perspectives* 118, no. 6 (2010): 884–89.

Graeber, David. *Direct Action: An Ethnography*. Oakland, CA: AK Press, 2009.

Graham, David. "History of *Helicobacter Pylori*, Duodenal Ulcer, Gastric Ulcer and Gastric Cancer." *World Journal of Gastroenterology* 20, no. 18 (2014): 5191–204.

Graham, Stephen, ed. *Disrupted Cities: When Infrastructure Fails*. New York: Routledge, 2010.

———. "Lessons in Urbicide." *New Left Review* 19 (2003): 63–77.

Graham, Stephen, and Simon Marvin. *Splintering Urbanism: Networked Infrastructures, Technological Mobilities and the Urban Condition*. New York: Routledge, 2001.

Grainger, Matt, and Kate Geary. "The New Forests Company and Its Ugandan Plantations." Oxfam and Uganda Land Alliance, 2011. https://www.oxfam.org/en/research/new-forests-company-and-its-uganda-plantations-oxfam-case-study.

Gray, Herman. "Introduction: Subject to Respectability." *Souls* 18, no. 2–4 (2016): 192–200.

Graziano, Valeria, and Kim Trogal. "Repair Matters." *Ephemera* 19, no. 2 (2019): 203–28.

Green, Elliott. "Decentralization and Development in Contemporary Uganda." *Regional & Federal Studies* 25, no. 5 (2015): 491–508.

———. "Patronage, District Creation, and Reform in Uganda." *Studies in Comparative International Development* 45, no. 1 (2010): 83–103.

Green-Simms, Lindsey. *Postcolonial Automobility: Car Culture in West Africa*. Minneapolis: University of Minnesota Press, 2017.

Gregson, Nicky, and Mike Crang. "From Waste to Resource: The Trade in Wastes and Global Recycling Economies." *Annual Review of Environment and Resources* 40, no. 1 (2015): 151–76.

———. "Materiality and Waste: Inorganic Vitality in a Networked World." *Environment and Planning A* 42, no. 2 (2010): 1026–32.

Gregson, Nicky, Michael Crang, F. Ahamed, N. Akhtar, and R. Ferdous. "Following Things of Rubbish Value: End-of-Life Ships, 'Chock-Chocky' Furniture and the Bangladeshi Middle Class Consumer." *Geoforum*. 41 (2010): 846–54.

Gregson, Nicky, Alan Metcalfe, and Louise Crewe. "Practices of Object Maintenance and Repair: How Consumers Attend to Consumer Objects within the Home." *Journal of Consumer Culture* 9, no. 2 (2009): 248–72.

The Guardian. "Ugandan Farmers Take on Palm Oil Giants over Land Grab Claims." March 3, 2015. https://www.theguardian.com/global-development/2015/mar/03/ugandan-farmers-take-on-palm-oil-giants-over-land-grab-claims

———. "Ugandan Opposition Leader Temporarily Blinded in Teargas Raid." April 28, 2011. https://www.theguardian.com/world/2011/apr/28/uganda-police-teargas-arrest-opposition-leader.

Gunston, Henry. "The Planning and Construction of the Uganda Railway." *Transactions of the Newcomen Society* 74, no. 1 (2004): 45–71.

Gupta, Akhil. "Suspension." *Theorizing the Contemporary: Fieldsites (Cultural Anthropology)* (blog), September 24, 2015. https://culanth.org/fieldsights/suspension.

Gutkind, Peter. *The Royal Capital of Buganda: A Study of Internal Conflict and External Ambiguity.* The Hague: Mouton & Co, 1963.

Gutschow, Kai. "Das Neue Afrika: Ernst May's 1947 Kampala Plan as Cultural Program." In *Colonial Architecture and Urbanism in Africa: Intertwined and Contested Histories,* edited by Fassil Demissie, 236–68. London: Ashgate, 2009.

———. "Modern Planning as Civilizing Agent: Ernst May's Kampala Extension Scheme." In *Proceedings of the 91st ASCA Annual Meeting,* 240–48, 2004.

Hage, Ghassan. *Alter-Politics: Critical Anthropology and the Radical Imagination.* Melbourne: Melbourne University Publishing, 2015.

Hansen, Holger, and Michael Twaddle, eds. *Changing Uganda: The Dilemmas of Structural Adjustment and Revolutionary Change.* Athens: Ohio University Press, 1991.

Hansen, Karen, ed. *African Encounters with Domesticity.* New Brunswick, NJ: Rutgers University Press, 1992.

Hansen, Karen, and Mariken Vaa, eds. *Reconsidering Informality: Perspectives from Urban Africa.* Uppsala, Sweden: Nordic Africa Institute, 2004.

Hanson, Holly. *Landed Obligation: The Practice of Power in Buganda.* Portsmouth, NH: Heinemann, 2003.

Haraway, Donna. *When Species Meet.* Minneapolis: University of Minnesota Press, 2008.

Harms, Erik. *Luxury and Rubble: Civility and Dispossession in the New Saigon.* Oakland: University of California Press, 2016.

Harvey, Penny, and Hannah Knox. "The Enchantments of Infrastructure." *Mobilities* 7, no. 4 (2012): 521–36.

Hawkins, Gay. *The Ethics of Waste: How We Relate to Rubbish.* Lanham, MD: Rowan & Littlefield, 2006.

Heidegger, Martin. *Being and Time.* New York: Harper Perennial, 1962.

Heron, Barbara. *Desire for Development: Whiteness, Gender and the Helping Imperative.* Waterloo, ON, Canada: Wilfred Laurier University Press, 2007.

Hestvik, Elin, Thorkild Tylleskar, Grace Ndeezi, Lena Grahnquist, Edda Olafsdottir, James K. Tumwine, and Deogratias H. Kaddu-Mulindwa. "Prevalence of *Helicobacter Pylori* in HIV-Infected, HAART-Naïve Ugandan Children: A Hospital-Based Survey." *Journal of the International AIDS Society* 14 (2011): 34.

Hetherington, Kregg. *Infrastructure, Environment, and Life in the Anthropocene.* Durham, NC: Duke University Press, 2019.

———. "Secondhandedness: Consumption, Disposal, and Absent Presence." *Environment and Planning D* 22, no. 1 (2004): 157–73.

Hickey, Sam, and Andries du Toit. "Adverse Incorporation, Social Exclusion, and Chronic Poverty." In *Chronic Poverty: Concepts, Causes and Policy,* edited by Andrew Shepherd and Julia Brunt, 134–59. Rethinking International Development Series. London: Palgrave Macmillan UK, 2013.

Higginbotham, Evelyn Brooks. *Righteous Discontent: The Women's Movement in the Black Baptist Church, 1880–1920.* Cambridge, MA: Harvard University Press, 1993.

Higgins, Maeve. "The Essential Workers America Treats as Disposable." *New York Review of Books* (blog), April 27, 2020. https://www.nybooks.com/daily/2020/04/27/the-essential-workers-america-treats-as-disposable.

Hird, Myra J. *The Origins of Sociable Life: Evolution after Science Studies.* Basingstoke, UK: Palgrave Macmillan, 2009.

Holmes, Douglas, and George Marcus. "Fast Capitalism: Para-Ethnography and the Rise of the Symbolic Analyst." In *Frontiers of Capital: Ethnographic Reflections on the New Economy*, edited by Melissa Fisher and Greg Downey, 33–56. Durham, NC: Duke University Press, 2006.

Hoornweg, Daniel, and Perinaz Bhaba-Tata. *What a Waste: A Global Review of Solid Waste Management.* Urban Development. Washington, DC: World Bank, 2012. https://openknowledge.worldbank.org/handle/10986/17388.

Hoy, Suellen. *Chasing Dirt: The American Pursuit of Cleanliness.* New York: Oxford University Press, 1995.

Human Rights Watch. *"Letting the Big Fish Swim": Failures to Prosecute High-Level Corruption in Uganda.* New York: Human Rights Watch, 2013. https://www.hrw.org/report/2013/10/21/letting-big-fish-swim/failures-prosecute-high-level-corruption-uganda.

———. "Uganda: Launch Independent Inquiry into Killings," May 8, 2011. https://www.hrw.org/news/2011/05/08/uganda-launch-independent-inquiry-killings.

———. "Uganda: Stop Harassing the Media." Human Rights Watch, May 20, 2013. https://www.hrw.org/news/2013/05/20/uganda-stop-harassing-media

Hundle, Anneeth Kaur. "African Asians and South Asians in Neoliberal Uganda: Culture, History and Political Economy." In *Uganda: The Dynamics of Neoliberal Transformation*, edited by Jorg Wiegratz, Giuliano Martiniello, and Elisa Greco, 285–302. London: Zed Books, 2018.

———. "Insecurities of Expulsion: Emergent Citizenship Formations and Political Practices in Postcolonial Uganda." *Comparative Studies of South Asia, Africa and the Middle East* 39, no. 1, (2019): 8–23.

Hurston, Zora Neale. "How It Feels to Be Colored Me." In *I Love Myself When I Am Laughing . . . and Then Again When I Am Looking Mean and Impressive: A Zora Neale Hurston Reader*, 152–55. Old Westbury, NY: Feminist Press, 1979.

iFixit. "Self-Repair Manifesto." July 21, 2020. https://www.ifixit.com/Manifesto.

Ignatiev, Noel. *How the Irish Became White.* New York: Routledge, 1995.

The Independent. "How Museveni Killed Anti-Lukwago Bill." Uganda, November 30, 2015.

———. "Jennifer Musisi's Kampala." May 4, 2014.

———. "Kampala City Commuter Train Service." March 15, 2015.

———. "Kampala's Lord Mayor Lukwago." March 18, 2011.

———. "Remembering the KCCA Evictions." October 4, 2015.

In Kampala. "City Yange—A Clean Sweep." August 31, 2012. https://www
.inkampala.com/news/city-yange-a-clean-sweep.

Izama, Angelo. "Is Yoweri Museveni Still the West's Man in Africa?" *African
Arguments* (blog), March 4, 2014. https://africanarguments.org/2014/03/is
-yoweri-museveni-still-the-wests-man-in-africa-by-angelo-izama.

———. "Kony Is Not the Problem." *New York Times.* March 20, 2012. https://
www.nytimes.com/2012/03/21/opinion/in-uganda-kony-is-not-the-only
-problem.html.

Izama, Angelo, and Edward Echwalu. "Season of Dissent." *Transition* 106
(2011): b58–71.

Izama, Angelo, and Mike Wilkerson. "Uganda: Museveni's Triumph and
Weakness." *Journal of Democracy* 22, no. 3 (2011): 64–78.

Jackson, Steven. "Rethinking Repair." In *Media Technologies: Essays on
Communication, Materiality, and Society,* edited by Tarleton Gillespie,
Pablo Boczkowski, and Kristen Foot. Cambridge, MA: MIT Press, 2014.

John van Nostrand Associates. *Kampala Urban Study: Phase One Report,* June
1993.

Joyce, Patrick. *The Rule of Freedom: Liberalism and the Modern City.* New
York: Verso, 2003.

Kagoro, Jude. "Competitive Authoritarianism in Uganda: The Not So Hidden
Hand of the Military." *Zeitschrift Für Vergleichende Politikwissenschaft* 10,
no. S1 (2016): 155–72.

Kampala Capital City Authority. "Kampala Physical Development Plan."
Kampala: KCCA, September 2012. https://www.kcca.go.ug/uploads/KPDP
%20Draft%20Final%20Report.pdf.

———. "Shut Down of Operations by KCCA." Press release. November 8, 2013.

———. *Strategic Plan 2014/15–2018/19.* Kampala: KCCA, 2014. https://www.kcca
.go.ug/uDocs/KCCA-STRATEGIC-PLAN-2014-19.pdf.

Kampala City Council. "The Local Governments (Kampala City Council) (Solid
Waste Management) Ordinance." Republic of Uganda, 2000.

———. "Strategy to Improve Solid Waste Management in Kampala City
(1999/2002)." Programme Coordination Unit, 1999.

Kampala City Yange. "About Us." 2011.

———. "Awards." 2012.

Kamya, Beti. "Ugandans Are Poor Because They Are Lazy and They Celebrate
Mediocrity." *Daily Monitor.* October 23, 2014.

Karlström, Mikael. "Imagining Democracy: Political Culture and Democratisa-
tion in Buganda." *Africa* 66, no. 4 (1996): 485–505.

———. "Modernity and Its Aspirants: Moral Community and Developmental
Eutopianism in Buganda." *Current Anthropology* 45, no. 5 (2004): 595–619.

Kasfir, Nelson. "State, Magendo, and Class Formation in Uganda." *Journal of Commonwealth and Comparative Politics* 21, no. 3 (1983): 84–103.

Kasozi, A. B. K. *The Social Origins of Violence in Uganda, 1964–1985.* Buffalo, NY: McGill-Queen's University Press, 1994.

Katusiimeh, Mesharch, Kees Burger, and Arthur Mol. "Informal Waste Collection and Its Co-Existence with the Formal Waste Sector: The Case of Kampala, Uganda." *Habitat International* 38 (2013): 1–9.

Katusiimeh, Mesharch, and Arthur Mol. "Environmental Legacies of Major Events: Solid Waste Management and the Commonwealth Heads of Government Meeting (CHOGM) in Uganda." *African Studies Quarterly* 12, no. 3 (2011): 47–65.

Katusiimeh, Mesharch, Arthur Mol, and Kees Burger. "The Operations and Effectiveness of Public and Private Provision of Solid Waste Collection Services in Kampala." *Habitat International* 36, no. 2 (2012): 247–52.

Kelley, Robin D. G. "'We Are Not What We Seem': Rethinking Black Working-Class Opposition in the Jim Crow South." *Journal of American History* 80, no. 1 (1993): 75–112.

Kennedy, Greg. *An Ontology of Trash: The Disposable and Its Problematic Nature.* Albany, NY: State University of New York Press, 2008.

Kibala Bauer, George. "Beyond the Western Gaze." *Africa Is a Country* (blog), May 29, 2020. https://africasacountry.com/2020/05/beyond-the-western-gaze.

Kigongo, Remigius, and Andrew Reid. "Local Communities, Politics and the Management of the Kasubi Tombs, Uganda." *World Archaeology* 39, no. 3 (2007): 371–84.

Kimari, Wangui, and Henrik Ernstson. "Imperial Remains and Imperial Invitations: Centering Race within the Contemporary Large-Scale Infrastructures of East Africa." *Antipode* 52, no. 3 (2020): 825–46.

Kinyanjui, Mary Njeri. *Women and the Informal Economy in Urban Africa: From the Margins to the Centre.* London: Zed Books, 2014.

Kirby, Peter Wynn. "Slow Burn: Dirt, Radiation, and Power in Fukushima." *Asia-Pacific Journal* 19, no. 3 (2019).

Knowles, Caroline. *Flip-Flop: A Journey through Globalisation's Backroads.* New York: Pluto Press, 2014.

Kodesh, Neil. *Beyond the Royal Gaze: Clanship and Public Healing in Buganda.* Charlottesville: University of Virginia Press, 2010.

Komakech, Allan, Noble Banadda, Joel Kinobe, Levi Kasisira, Cecilia Sundberg, Girma Gebresenbet, and Björn Vinnerås. "Characterization of Municipal Waste in Kampala, Uganda." *Journal of the Air & Waste Management Association* 64, no. 3 (2014): 340–48.

Kooy, Michelle, and Karen Bakker. "Splintered Networks: The Colonial and Contemporary Waters of Jakarta." *Geoforum* 39, no. 6 (2008): 1843–58.

Kotef, Hagar. *Movement and the Ordering of Freedom: On Liberal Governances of Mobility.* Durham, NC: Duke University Press, 2015.

Kothari, Uma. "An Agenda for Thinking about 'Race' in Development." *Progress in Development Studies* 6, no. 1 (2006): 9–23.

Kuletz, Valerie. *The Tainted Desert: Environmental Ruin in the American West.* New York: Routledge, 1998.

Kyomuhendo, Grace Bantebya, and Marjorie McIntosh. *Women, Work and Domestic Virtue in Uganda, 1900–2003.* Athens: Ohio University Press, 2006.

Lakoff, Andrew, and Stephen Collier. "Infrastructure and Event: The Political Technology of Preparedness." In *Political Matter: Technoscience, Democracy, and Public Life,* edited by Bruce Braun and Sarah Whatmore, 243–66. Minneapolis: University of Minnesota Press, 2010.

Lambright, Gina. "Opposition Politics and Urban Service Delivery in Kampala, Uganda." *Development Policy Review* 32, no. s1 (2014): s39–60.

Laporte, Dominique. *History of Shit.* Cambridge, MA: MIT Press, 2002.

Larkin, Brian. "The Politics and Poetics of Infrastructure." *Annual Review of Anthropology* 42 (2013): 327–43.

———. "Promising Forms: The Political Aesthetics of Infrastructure." In *The Promise of Infrastructure,* edited by Nikhil Anand, Akhil Gupta, and Hannah Appel, 175–202. Durham, NC: Duke University Press, 2018.

Laruni, Elizabeth. "From the Village to Entebbe: The Acholi of Northern Uganda and the Politics of Identity, 1950–1985." PhD. Diss. University of Exeter, 2014.

Law, John. *Organizing Modernity.* Cambridge, MA: Blackwell, 1994.

Lawhon, Mary, David Nilsson, Jonathan Silver, Henrik Ernstson, and Shuaib Lwasa. "Thinking through Heterogeneous Infrastructure Configurations." *Urban Studies,* 2017.

Lee, Hedwig, and Margaret Takako Hicken. "Death by a Thousand Cuts: The Health Implications of Black Respectability Politics." *Souls* 18, no. 2–4 (2016): 421–45.

Leonard, Lori. *Life in the Time of Oil: A Pipeline and Poverty in Chad.* Bloomington: Indiana University Press, 2016.

Lerner, Steve. *Sacrifice Zones: The Front Lines of Toxic Chemical Exposure in the United States.* Cambridge, MA: MIT Press, 2010.

Li, Tania Murray. *Land's End: Capitalist Relations on an Indigenous Frontier.* Durham, NC: Duke University Press, 2014.

———. *The Will to Improve: Governmentality, Development, and the Practice of Politics.* Durham, NC: Duke University Press, 2007.

Liboiron, Max. "How Plastic Is a Function of Colonialism." *Teen Vogue,* 2018. https://www.teenvogue.com/story/how-plastic-is-a-function-of-colonialism.

———. *Pollution Is Colonialism.* Durham, NC: Duke University Press, 2021.

———. "Recycling as a Crisis of Meaning." *E-Topia* 4 (2009).

———. "Waste Colonialism." *Discard Studies* (blog), November 1, 2018. https://discardstudies.com/2018/11/01/waste-colonialism.

———. "Waste Is Not 'Matter Out of Place.'" *Discard Studies* (blog), September 9, 2019. https://discardstudies.com/2019/09/09/waste-is-not-matter-out-of-place.

———. "Why Discard Studies?" *Discard Studies*, May 7, 2014. https://discardstudies.com/2014/05/07/why-discard-studies.

Lie, Jon Harald Sande. *Developmentality: An Ethnography of the World Bank–Uganda Partnership.* New York: Berghahn Books, 2015.

Lindell, Ilda, ed. *Africa's Informal Workers: Collective Agency, Alliances and Transnational Organizing in Urban Africa.* London: Zed Books, 2010.

Lindell, Ilda, and Jenny Appelblad. "Disabling Governance: Privatisation of City Markets and Implications for Vendors' Associations in Kampala, Uganda." *Habitat International* 33 (2009): 397–404.

Livingston, Julie. *Improvising Medicine: An African Oncology Ward in an Emerging Cancer Epidemic.* Durham, NC: Duke University Press, 2012.

———. *Self-Devouring Growth: A Planetary Parable as Told from Southern Africa.* Durham, NC: Duke University Press, 2019.

Locke, John. *Two Treatises of Government.* Cambridge Texts in the History of Political Thought. New York: Cambridge University Press, 1960.

Loftsdóttir, Kristín. "Invisible Colour: Landscapes of Whiteness and Racial Identity in International Development." *Anthropology Today* 25, no. 5 (2009): 4–7.

Lorimer, Jamie. "Parasites, Ghosts and Mutualists: A Relational Geography of Microbes for Global Health." *Transactions of the Institute of British Geographers* 42, no. 4 (2017): 544–58.

Low, Anthony. "Religion and Society in Buganda 1875–1900." *East African Studies* 8 (1957): 1–16.

Lucas, Gavin. "Disposability and Dispossession in the Twentieth Century." *Journal of Material Culture* 7, no. 1 (2002): 5–22.

MacBride, Samantha. *Recycling Reconsidered: The Present Failure and Future Promise of Environmental Action in the United States.* Cambridge, MA: MIT Press, 2011.

Macpherson, C. B. *The Political Theory of Possessive Individualism: Hobbes to Locke.* New York: Oxford University Press, 1962.

Magubane, Bernard. *Race and the Construction of the Dispensable Other.* Pretoria: University of South Africa Press, 2007.

Mains, Daniel. "Blackouts and Progress: Privatization, Infrastructure, and a Developmentalist State in Jimma." *Cultural Anthropology* 27, no. 1 (2012): 3–27.

———. *Under Construction: Technologies of Development in Urban Ethiopia.* Durham, NC: Duke University Press, 2019.

Makhulu, Anne-Maria. *Making Freedom: Apartheid, Squatter Politics, and the Struggle for Home*. Durham, NC: Duke University Press, 2015.

Malkki, Liisa. *The Need to Help: The Domestic Arts of International Humanitarianism*. Durham, NC: Duke University Press, 2015.

Mamdani, Mahmood. *Imperialism and Fascism in Uganda*. London: Heinemann Educational Books, 1983.

———. *Politics and Class Formation in Uganda*. New York: Monthly Review Press, 1976.

———. "The Ugandan Asian Expulsion: Twenty Years After." *Journal of Refugee Studies* 6, no. 3 (1993): 265–73.

———. *When Victims Become Killers: Colonialism, Nativism, and the Genocide in Rwanda*. Princeton, NJ: Princeton University Press, 2001.

Mann, Michael. "The Autonomous Power of the State: Its Origins, Mechanisms and Results." *European Journal of Sociology / Archives Européennes de Sociologie* 25, no. 2 (1984): 185–213.

Marx, Karl. *The Eighteenth Brumaire of Louis Bonaparte*. New York: International Publishers, 1852.

Masco, Joseph. "'Survival Is Your Business': Engineering Ruins and Affect in Nuclear America." *Cultural Anthropology* 23, no. 2 (2008): 361–98.

Massumi, Brian. "The Autonomy of Affect." *Cultural Critique* 31 (1995): 83–109.

Matlon, Jordanna. "This Is How We Roll: The Status Economy of Bus Portraiture in the Black Urban Periphery." *Laboratorium* 7, no. 2 (2015): 62–82.

Mattern, Shannon. "Maintenance and Care." *Places Journal*, 2018. https://placesjournal.org/article/maintenance-and-care.

Mazzarella, William. "Affect: What Is It Good For?" In *Enchantments of Modernity: Empire, Nation, Globalization*, edited by Saurabh Dube, 291–309. New York: Routledge, 2009.

Mbembe, Achille. "At the Edge of the World: Boundaries, Territoriality, and Sovereignty in Africa." *Public Culture* 12, no. 1 (2000): 259–84.

———. "Necropolitics." *Public Culture* 15, no. 1 (2003): 11–40.

———. *On the Postcolony*. Berkeley: University of California Press, 2001.

Mbembe, Achille, and Janet Roitman. "Figures of the Subject in Times of Crisis." *Public Culture* 7, no. 2 (1995): 323–52.

McClintock, Anne. *Imperial Leather: Race, Gender and Sexuality in the Colonial Contest*. New York: Routledge, 1995.

McFarlane, Colin, and Jonathan Silver. "Navigating the City: Dialectics of Everyday Urbanism." *Transactions of the Institute of British Geographers* 42, no. 3 (2017): 458–71.

Meagher, Kate. "Cannibalizing the Informal Economy: Frugal Innovation and Economic Inclusion in Africa." *European Journal of Development Research* 30, no. 1 (2018): 17–33.

Meagher, Kate, and Ilda Lindell. "Engaging with African Informal Economies: Social Inclusion or Adverse Incorporation?" *African Studies Review* 56, no. 3 (2013): 57–76.

Mehta, Uday. *Liberalism and Empire: A Study in Nineteenth-Century British Liberal Thought.* Chicago: University of Chicago Press, 1999.

Melly, Caroline. *Bottleneck: Mobility, Building, and Belonging in an African City.* Chicago: University of Chicago Press, 2017.

Melosi, Martin. *Garbage in the Cities: Refuse Reform and the Environment.* Pittsburgh: University of Pittsburgh Press, 2004.

Millar, Kathleen. "Garbage as Racialization." *Anthropology and Humanism* 45, no. 1 (2020): 4–24.

———. "The Precarious Present: Wageless Labor and Disrupted Life in Rio de Janeiro, Brazil." *Cultural Anthropology* 29, no. 1 (2014): 32–53.

———. *Reclaiming the Discarded: Life and Labor on Rio's Garbage Dump.* Durham, NC: Duke University Press, 2018.

Minter, Adam. *Junkyard Planet: Travels in the Billion-Dollar Trash Trade.* New York: Bloomsbury, 2013.

Miraftab, Faranak. "Neoliberalism and Casualization of Public Sector Services: The Case of Waste Collection Services in Cape Town, South Africa." *International Journal of Urban and Regional Research* 28, no. 4 (2004): 874–92.

Mirams, A. E. *Kampala: Report on the Town Planning and Development of Kampala.* Entebbe, Uganda: Government Printer, 1930.

Mitchell, Timothy. *Rule of Experts: Egypt, Techno-Politics, Modernity.* Berkeley: University of California Press, 2002.

Le Monde Diplomatique. "Cheap Help from Uganda." May 1, 2012. https://mondediplo.com/2012/05/05uganda.

Monteith, William. "Markets and Monarchs: Indigenous Urbanism in Postcolonial Kampala." *Settler Colonial Studies* 9, no. 2 (2019): 247–65.

Moore, Erin. "Idle Women, Bored Sex, and the Temporal Interventions of Population Control." *Voices* 13, no. 1 (2018): 74–87.

———. "Postures of Empowerment: Cultivating Aspirant Feminism in a Ugandan NGO." *Ethos* 44, no. 3 (2016): 375–96.

———. "What the Miniskirt Reveals: Sex Panics, Women's Rights, and Pulling Teeth in Urban Uganda." *Anthropological Quarterly* 93, no. 3 (2020): 321–350.

Morris, Susana. *Close Kin and Distant Relatives: The Paradox of Respectability in Black Women's Literature.* Charlottesville: University of Virginia Press, 2014.

Morton, David. *Age of Concrete: Housing and the Shape of Aspiration in the Capital of Mozambique.* Athens: Ohio University Press, 2019.

Mugaga, Frank. "The Public–Private Sector Approach to Municipal Solid Waste Management. How Does It Work in Makindye Division, Kampala District, Uganda?" MA thesis, Norwegian University of Science and Technology, 2006.

Muir, Sarah, and Akhil Gupta. "Rethinking the Anthropology of Corruption: An Introduction to Supplement 18." *Current Anthropology* 59, no. S18 (2018): S4–15.

Mukasa, Ham, John A. Rowe, and Andrew Kabazza. *Simuda Nyuma: Ebiro Bya Mutesa = Do Not Go Back : The Reign of Mutesa*. Kampala, 1962.

Mukherjee, Rahul. *Radiant Infrastructures: Media, Environment, and Cultures of Uncertainty*. Durham, NC: Duke University Press, 2020.

Mukiibi, Stephen. "Policies on the Supply of Housing to Urban Low-Income Earners in Uganda: A Case Study of Kampala." PhD diss., Newcastle University, 2008.

Mukwaya, Paul, Hannington Sengendo, and Shuaib Lwasa. "Urban Development Transitions and Their Implications for Poverty Reduction and Policy Planning in Uganda." WIDER Working Paper 2010/045. Helsinki: UNU-WIDER, 2010. https://www.wider.unu.edu/publication/urban-development-transitions-and-their-implications-poverty-reduction-and-policy.

Murphy, Michelle. "Chemical Infrastructures of the St Clair River." In *Toxicants, Health and Regulation since 1945*, edited by Soraya Boudia and Nathalie Jas, 103–15. London: Pickering and Chato, 2013.

———. "Chemical Regimes of Living." *Environmental History* 13, no. 4 (2008): 695–703.

Museveni, Yoweri. "Fanon's Theory of Violence, Its Verification in Liberated Mozambique." In *Essays on the Liberation of Southern Africa*, edited by Nathan Shamuyarira, 1–24. Dar es Salaam, Tanzania: Tanzania Publishing House, 1971.

——— (@KagutaMuseveni). "The old man with the hat." Twitter, January 4, 2019, 1:22 p.m. https://twitter.com/kagutamuseveni/status/1081254657132036098?lang=en.

———. *Sowing the Mustard Seed: The Struggle for Freedom and Democracy in Uganda*. Kampala: Moran Publishers, 1997.

Musisi, J. Semakula. "Statement on the Issues of Kampala Capital City Authority." A letter to Parliament Sectoral Committee on Public Service and Local Government on behalf of KCCA, February 6, 2013. https://www.kcca.go.ug/uploads/Statement.pdf.

Musisi, Nakanyike. "Colonial and Missionary Education: Women and Domesticity in Uganda, 1900–1945." In *African Encounters with Domesticity*, edited by Karen Tranberg Hansen, 172–94. New Brunswick, NJ: Rutgers University Press, 1992.

Mutambatsere, E., A. Nalikka, M. Pal, and D. Vencatachellum. "What Role for Multilateral Development Banks in Project Finance? Some Thoughts from the Rift Valley Railways in Kenya and Uganda." *Journal of Infrastructure Development* 5, no. 1 (2013): 1–20.

Mwenda, Andrew. "How to Change Kampala." *The Independent*, Uganda, May 13, 2012. https://www.independent.co.ug/change-kampala-part-1.

———. "Personalizing Power in Uganda." *Journal of Democracy* 18, no. 3 (2007): 23–37.

Mwenda, Andrew, and Roger Tangri. "Patronage Politics, Donor Reforms, and Regime Consolidation in Uganda." *African Affairs* 104, no. 416 (2005): 449–67.

Mwiganga, Moses, and Frank Kasiime. "The Impact of Mpererwe Landfill in Kampala–Uganda, on the Surrounding Environment." *Physics and Chemistry of the Earth, Parts A/B/C* 30, nos. 11–16 (2005): 74–750.

Myers, Garth. *Disposable Cities: Garbage, Governance and Sustainable Development in Urban Africa*. Burlington, VT: Ashgate, 2005.

———. *Rethinking Urbanism: Lessons from Postcolonialism and the Global South*. Bristol, UK: Bristol University Press, 2020.

———. *Verandahs of Power: Colonialism and Space in Urban Africa*. Syracuse, NY: Syracuse University Press, 2003.

Naggaga, William. "Were African Countries Ready for Independence?" *Daily Monitor*. October 26, 2018.

Nagle, Robin. *Picking Up: On the Streets and Behind the Trucks with the Sanitation Workers of New York City*. New York: Farrar, Straus and Giroux, 2014.

Nagy, Kelsi, and Phillip David Johnson, eds. *Trash Animals: How We Live with Nature's Filthy, Feral, Invasive, and Unwanted Species*. Minneapolis: University of Minnesota Press, 2013.

Namara, Agrippinah. "The Invisible Workers: Paid Domestic Work in Kampala City, Uganda." Kampala: Center for Basic Research, 2001.

Nation, Mark Thiessen. "Footwashing: Preparation for Christian Life." In *The Blackwell Companion to Christian Ethics*, edited by Stanley Hauerwas and Samuel Wells, 479–90. Malden, MA: Blackwell, 2011.

National Association of Professional Environmentalists. *A Study on Land Grabbing Cases in Uganda*. Kampala: National Association of Professional Environmentalists, 2012. https://reliefweb.int/sites/reliefweb.int/files/resources/Full_Report_3823.pdf.

National Environmental Management Authority. "Draft Guidelines for Solid Waste Management in Uganda." Republic of Uganda, 2003.

Ndlovu-Gatsheni, Sabelo. "Coloniality of Power in Development Studies and the Impact of Global Imperial Designs on Africa." *ARAS* 33, no. 2 (2012): 48–73.

New Vision. "Budget: Experts React." July 13, 2013.

———. "Face to Face with Jennifer Musisi." May 5, 2012.

———. "For the Sins of Uganda, I Repent—Museveni." October 18, 2012.

———. "Four Killed in Kampala Floods." November 17, 2007.

———. "Germany Halts Aid to Uganda over Corruption Scandal." December 2, 2012.

———. "I Am Tired of Cheap Politicking—Museveni." January 14, 2013.

———. "Kampala a Mountain of Garbage." November 19, 2008.

———. "Kampala Is Better Now without Lukwago." June 27, 2015.

———. "Kampala on Its Way to Being a Clean City." September 12, 2011.

———. "KCCA Demolishes Kiosks in Old and New Taxi Parks." May 10, 2012.

———. "KCCA Employees Back to Work." December 1, 2013.

———. "KCCA to Introduce Railway Service to Ease Jam." February 2, 2012.

———. "Kerosene Tax Angers Female MPs." June 28, 2013.

———. "Lazy, Thieving Ugandans Give Jobs to Foreigners." October 26, 2013.

———. "Lazy Workforce Derails Growth, Says Museveni." May 1, 2014.

———. "Museveni Explains Paraffin, Water Taxes." July 16, 2013.

———. "Museveni Orders City Drainages Fixed." November 19, 2007.

———. "Nakawa-Naguru Tenants Evicted." July 5, 2011.

———. "One Year on—Jennifer Musisi." April 30, 2012.

———. "OPM: World Bank Reassess Aid to Uganda." November 14, 2012.

———. "Pictorial: Kampala City Yange Cleans Bwaise." August 18, 2012.

———. "Rubaga Leaders Want Railway Evictees Compensated." August 10, 2014.

———. "We Are All Responsible for Garbage in the City." May 10, 2012.

———. "Who Is Responsible?" February 11, 2013.

Newell, Sasha. *The Modernity Bluff: Crime, Consumption, and Citizenship in Côte d'Ivoire.* Chicago: University of Chicago Press, 2012.

Newell, Stephanie. *Histories of Dirt: Media and Urban Life in Colonial and Postcolonial Lagos.* Durham, NC: Duke University Press, 2020.

New York Times. "Michael Brown Spent Last Weeks Grappling with Problems and Promise." August 24, 2014, sec. U.S. https://www.nytimes.com/2014/08/25/us/michael-brown-spent-last-weeks-grappling-with-lifes-mysteries.html.

Nguyen, Minh. *Waste and Wealth: An Ethnography of Labor, Value, and Morality in a Vietnamese Recycling Economy.* Oxford: Oxford University Press, 2018.

Niringiye, Aggrey, and Omortor Douglason. "Determinants of Willingness to Pay for Solid Waste Management in Kampala City." *Current Research Journal of Economic Theory* 2, no. 3 (2010): 119–22.

Nixon, Rob. *Slow Violence and the Environmentalism of the Poor.* Cambridge, MA: Harvard University Press, 2011.

Nkurunziza, Emmanuel. "Informal Mechanisms for Accessing and Securing Urban Land Rights: The Case of Kampala, Uganda." *Environment and Urbanization* 19, no. 2 (2007): 509–26.

———. "Two States, One City? Conflict and Accommodation in Land Delivery in Kampala, Uganda." *International Development Planning Review* 28, no. 2 (2006): 159–80.

Nunzio, Marco Di. *The Act of Living: Street Life, Marginality, and Development in Urban Ethiopia*. Ithaca, NY: Cornell University Press, 2019.

Nuwagaba, Augustus. "Dualism in Kampala: Squalid Slums in a Royal Realm." In *African Urban Economies: Viability, Vitality or Vitiation?*, edited by Deborah Bryceson and Deborah Potts, 151–67. New York: Palgrave Macmillan, 2006.

———. "The Impact of Macro-Adjustment Programmes on Housing Investment in Kampala City—Uganda: Shelter Implications for the Urban Poor." *East African Social Science Research Review* 16, no. 1 (2000): 49–64.

Nyanzi, Stella. *No Roses from My Mouth: Poems from Prison*. Ubuntu Reading Group, 2020.

Nye, David. *American Technological Sublime*. Cambridge, MA: MIT Press, 1994.

Obbo, Christine. *African Women: Their Struggle for Economic Independence*. London: Zed Books, 1980.

———. "Women, Children and a 'Living Wage'." In *Changing Uganda: The Dilemmas of Structural Adjustment and Revolutionary Change*, edited by Holger Hansen and Michael Twaddle, 98–112. Kampala: Fountain Publishers, 1991.

Obita, Julius. "Justice in Land: Assessing the Impact on the Evicted Tenants of Nsambya Railway Quarters, Kampala." Master's thesis, Norwegian University of Science and Technology, 2016.

The Observer. "City Dwellers Get Brooms as KCCA Strives to Make Garbage History." Uganda, March 29, 2012.

———. "KCCA at 1 Year: Fights Continue to Rock KCCA." March 14, 2012.

———. "Museveni's Four-Acre Prosperity Formula Is Possible." June 25, 2014.

———. "Musisi Wants Lukwago to Have 'Eyes on Hands off' Role." March 27, 2012.

Ogden, Jessica. "'Producing' Respect: The 'Proper Woman' in Postcolonial Kampala." In *Postcolonial Identities in Africa*, edited by Richard Werbner and Terence Ranger, 165–92. London: Zed Books, 1996.

Oliver, Roland. "The Royal Tombs of Buganda." *Uganda Journal* 23, no. 2 (1959): 129–34.

Oloka-Onyango, Joseph. "New-Breed Leadership, Conflict, and Reconstruction in the Great Lakes Region of Africa: A Sociopolitical Biography of Uganda's Yoweri Kaguta Museveni." *Africa Today* 50, no. 3 (2004): 29–52.

Omi, Michael, and Howard Winant. *Racial Formation in the United States: From the 1960s to the 1990s*. New York: Routledge, 1994.

Omolo-Okalebo, Fredrick, Tigran Haas, Inga Britt Werner, and Hannington Sengendo. "Planning of Kampala City 1903–1962: The Planning Ideas, Values, and Their Physical Expression." *Journal of Planning History* 9, no. 3 (2010): 151–69.

The Onion. "6-Day Visit to Rural African Village Completely Changes Woman's Facebook Profile Picture." January 28, 2014. https://www.theonion.com/6-day-visit-to-rural-african-village-completely-changes-1819576037.

Oteng-Ababio, Martin, Jose Ernesto Melara Arguello, and Offira Gabbay. "Solid Waste Management in African Cities: Sorting the Facts from the Fads in Accra, Ghana." *Habitat International* 39 (2013): 96–104.

Otter, Christopher. "Cleansing and Clarifying: Technology and Perception in Nineteen-Century London." *Journal of British Studies* 43, no. 1 (2004): 40–64.

———. "Making Liberalism Durable: Vision and Civility in the Late Victorian City." *Social History* 27, no. 1 (2002): 1–15.

Packard, Vance. *The Wastemakers*. New York: LG Publishing, 1960.

Page, Max. *The City's End: Two Centuries of Fantasies, Fears, and Premonitions of New York's Destruction*. New Haven, CT: Yale University Press, 2008.

Pailey, Robtel Neajai. "De-Centering the 'White Gaze' of Development." *Development and Change* 51, no. 3 (2020): 729–45.

Parikh, Shanti. "From Auntie to Disco: The Bifurcation of Risk and Pleasure in Sex Education in Uganda." In *Sex in Development: Science, Sexuality, and Morality in Global Perspective*, edited by Vincanne Adams and Stacy Leigh Pigg, 125–58. Durham, NC: Duke University Press, 2005.

Parkin, David. *Neighbors and Nationals in an African City Ward*. Berkeley: University of California Press, 1969.

Parnell, Sue, and Edgar Pieterse, eds. *Africa's Urban Revolution*. London: Zed Books, 2014.

Patel, Raj, and Jason Moore. *A History of the World in Seven Cheap Things: A Guide to Capitalism, Nature, and the Future of the Planet*. Oakland: University of California Press, 2017.

Paxson, Heather, and Stefan Helmreich. "The Perils and Promises of Microbial Abundance: Novel Natures and Model Ecosystems, from Artisanal Cheese to Alien Seas." *Social Studies of Science* 44, no. 2 (2013): 165–93.

Pellow, Deborah. *Landlords and Lodgers: Socio-Spatial Organization in an Accra Community*. Chicago: University of Chicago Press, 2008.

Pereira da Silva, Tatianna. "Follow the Bottle: PET Recycling Economy and Waste Picker Empowerment in Brazil." PhD diss., University of Edinburgh, 2020.

Philipps, Joschka, and Jude Kagoro. "The Metastable City and the Politics of Crystallisation: Protesting and Policing in Kampala." *Africa Spectrum* 51, no. 3 (2016): 3–32.

Pier, David. "The Branded Arena: Ugandan 'Traditional' Dance in the Marketing Era." *Africa: The Journal of the International African Institute* 81, no. 3 (2011): 413–33.

Pierce, Steven. "Looking Like a State: Colonialism and the Discourse of Corruption in Northern Nigeria." *Comparative Studies in Society and History* 48, no. 4 (2006): 887–914.

Pierre, Jemima. *The Predicament of Blackness: Postcolonial Ghana and the Politics of Race*. Chicago: University of Chicago Press, 2013.

——. "The Racial Vernaculars of Development: A View from West Africa." *American Anthropologist* 122, no. 1 (2020): 86–98.

——. "Slavery, Anthropological Knowledge, and the Racialization of Africans." *Current Anthropology* 61, no. S22 (2020): S220–S231.

——. "Structure, Project, Process: Anthropology, Colonialism, and Race in Africa." *Journal of Anthropological Sciences* 96 (2018): 213–19.

Pieterse, Edgar. "Grasping the Unknowable: Coming to Grips with African Urbanisms." *Social Dynamics* 37, no. 1 (2011): 5–23.

Povinelli, Elizabeth. *Economies of Abandonment: Social Belonging and Endurance in Late Liberalism.* Durham, NC: Duke University Press, 2011.

Prakash, Gyan. *Another Reason: Science and the Imagination of Modern India.* Princeton, NJ: Princeton University Press, 1999.

Pratt, Mary Louis. *Imperial Eyes: Travel Writing and Transculturation.* New York: Routledge, 1992.

Pulido, Laura. "A Critical Review of the Methodology of Environmental Racism Research." *Antipode* 28, no. 2 (1996): 142–59.

Quayson, Ato. *Oxford Street, Accra: City Life and the Itineraries of Transnationalism.* Durham, NC: Duke University Press, 2014.

Radio Netherlands Worldwide. "'People Threaten to Kill Me So Often, I've Gotten Used to It.'" November 25, 2011.

Raffles, Hugh. "Jews, Lice, and History." *Public Culture* 19, no. 3 (2007): 521–66.

Ranciere, Jacques. *The Philosopher and His Poor.* Durham, NC: Duke University Press, 2004.

Rao, Vyjayanthi. "How to Read a Bomb: Scenes from Bombay's Black Friday." *Public Culture* 19, no. 3 (2007): 567–92.

Rathje, William, and Murphey Cullen. *Rubbish! The Archaeology of Garbage.* Tucson: University of Arizona Press, 2001.

Ray, Benjamin. "Royal Shrines and Ceremonies of Buganda." *Uganda Journal* 36 (1972): 35–48.

Redfield, Peter, and Steven Robins. "An Index of Waste: Humanitarian Design, 'Dignified Living' and the Politics of Infrastructure in Cape Town." *Anthropology Southern Africa* 39, no. 2 (2016): 145–62.

RedPepper. "Govt: Railway Line Evictions to Continue." July 31, 2014.

Reid, Richard. *Political Power in Pre-Colonial Buganda.* Athens: Ohio University Press, 2002.

Reinert, Hugo. "Sacrifice." *Environmental Humanities* 7, no. 1 (2016): 255–58.

Reno, Joshua. "Waste and Waste Management." *Annual Review of Anthropology* 44, no. 1 (2015): 557–72.

——. *Waste Away: Working and Living with a North American Landfill.* Oakland: University of California Press, 2016.

———. "Your Trash Is Someone's Treasure: The Politics of Value at a Michigan Landfill." *Journal of Material Culture* 14, no. 1 (2009): 29–46.

Reporters without Borders. "Uganda—Two Kampala Newspapers Unable to Publish for Past Ten Days." Reporters Without Borders, May 29, 2013. https://news.trust.org/item/20130529091145-wfdpc/Bin It.

Republic of Uganda. "Constitution of the Republic of Uganda," 1995. https://www.statehouse.go.ug/government/constitution.

———. "The Free Zones Act, 2014," 2014. https://www.ilo.org/dyn/natlex/natlex4.detail?p_lang=en&p_isn=96870.

———. "Kampala Development Plan, Structure Report." Ministry of Provincial Administration, April 1972.

———. *Value for Money Audit Report on Solid Waste Management in Kampala.* Kampala: Office of the Auditor General, 2010. http://www.oag.go.ug/wp-content/uploads/2016/07/SOLID-WASTE.pdf.

Resnick, Danielle. *Urban Poverty and Party Populism in African Democracies.* New York: Cambridge University Press, 2014.

Reuters. "Three Die in Uganda as Floods Submerge Slum." November 16, 2007. https://www.reuters.com/article/uk-uganda-floods/three-die-in-uganda-as-floods-submerge-slum-idUKL1621186120071116?edition-redirect=uk.

Rhodes, Jane. "Pedagogies of Respectability: Race, Media, and Black Womanhood in the Early 20th Century." *Souls* 18, no. 2–4 (2016): 201–14.

Rizzo, Matteo. *Taken for a Ride: Grounding Neoliberalism, Precarious Labour, and Public Transport in an African Metropolis.* Oxford: Oxford University Press, 2017.

Rizzo, Matteo, Blandina Kilama, and Marc Wuyts. "The Invisibility of Wage Employment in Statistics on the Informal Economy in Africa: Causes and Consequences." *Journal of Development Studies* 51, no. 2 (2015): 149–61.

Robinson, Cedric. *Black Marxism: The Making of the Black Radical Tradition.* London: Zed Books, 1983.

Robinson, Jennifer. *Ordinary Cities: Between Modernity and Development.* New York: Routledge, 2013.

Rockson, George, Francis Kemausuor, Raymond Seassey, and Ernest Yanful. "Activities of Scavengers and Itinerant Buyers in Greater Accra, Ghana." *Habitat International* 39 (2013): 148–55.

Rodgers, Dennis, and Bruce O'Neill. "Infrastructural Violence: Introduction to the Special Issue." *Ethnography* 13, no. 4 (2012): 401–12.

Roediger, David. *Working toward Whiteness: How America's Immigrants Became White.* New York: Basic Book, 2005.

Rogers, Heather. *Gone Tomorrow: The Hidden Life of Garbage.* New York: Norton, 2005.

Rosaldo, Renato. *Culture and Truth: The Remaking of Social Analysis.* Boston: Beacon Press, 1989.

Ross, Fiona. *Raw Life, New Hope: Decency, Housing and Everyday Life in a Post-Apartheid Community.* Cape Town: University of Cape Town Press, 2010.

Roy, Ananya. *Poverty Capital: Microfinance and the Making of Development.* New York: Routledge, 2010.

———. "Who's Afraid of Postcolonial Theory?" *International Journal of Urban and Regional Research* 40, no. 1 (2016): 200–209.

Roy, Ananya, and Emma Shaw Crane, eds. *Territories of Poverty: Rethinking North and South.* Athens: University of Georgia Press, 2015.

Roy, Ananya, Genevieve Negrón-Gonzales, Kweku Opoku-Agyemang, and Clare Talwalker. *Encountering Poverty: Thinking and Acting in an Unequal World.* Oakland: University of California Press, 2016.

Roy, Ananya, Stuart Schrader, and Emma Shaw Crane. "'The Anti-Poverty Hoax': Development, Pacification, and the Making of Community in the Global 1960s." *Cities* 44 (2015): 139–45.

Royte, Elizabeth. *Garbage Land: On the Secret Trail of Trash.* New York: Little, Brown, 2007.

Rubongoya, Joshua. *Regime Hegemony in Museveni's Uganda: Pax Musevenica.* New York: Palgrave MacMillan, 2007.

Samarth, Ujwala. *The Occupational Health of Waste Pickers in Pune: KKPKP and SWaCH Members Push for Health Rights.* Cambridge, MA: WIEGO, 2014. https://www.wiego.org/sites/default/files/publications/files/Samarth_OHS_Health_of_WP_in_Pune.pdf.

Samson, Melanie. "Producing Privatization: Re-Articulating Race, Gender, Class and Space." *Antipode* 42, no. 2 (2010): 404–32.

———. "Rescaling the State, Restructuring Social Relations." *International Feminist Journal of Politics* 10, no. 1 (2008): 19–39.

Scanlan, John. *On Trash.* London: Reaktion Books, 2005.

Scherz, China. *Having People, Having Heart: Charity, Sustainable Development, and Problems of Dependence in Central Uganda.* Chicago: University of Chicago Press, 2014.

Schindler, Seth, and J. Miguel Kanai. "Getting the Territory Right: Infrastructure-Led Development and the Re-emergence of Spatial Planning Strategies." *Regional Studies* 55, no. 1 (2019): 1–12.

Schubeler, Peter, Karl Wehrle, and Jurg Christen. "Conceptual Framework for Municipal Solid Waste Management in Low-Income Countries." Urban Management and Infrastructure Working Papers. Washington, DC: World Bank, August 1, 1996. https://documents.worldbank.org/en/publication/documents-reports/documentdetail/829601468315304079/conceptual-framework-for-municipal-solid-waste-management-in-low-income-countries.

Schulz, Yvan, and Anna Lora-Wainwright. "In the Name of Circularity: Environmental Improvement and Business Slowdown in a Chinese Recycling Hub." *Worldwide Waste: Journal of Interdisciplinary Studies* 2, no. 1 (2019): 9.

Schumpeter, Joseph. *Capitalism, Socialism and Democracy*. New York: Routledge, 1976.

Scott, James. *Domination and the Arts of Resistance: Hidden Transcripts*. New Haven, CT: Yale University Press, 1990.

———. *Seeing Like a State: How Certain Schemes to Improve the Human Condition Have Failed*. New Haven, CT: Yale University Press, 1998.

Seattle Times. "Drugs Found in Puget Sound Salmon from Tainted Wastewater." February 23, 2016. https://www.seattletimes.com/seattle-news/environment/drugs-flooding-into-puget-sound-and-its-salmon.

Sedgwick, Eve Kosofsky. *Touching Feeling: Affect, Pedagogy, Performativity*. Durham, NC: Duke University Press, 2003.

Semboja, Joseph, and Ole Therkildsen, eds. *Service Provision under Stress in East Africa: The State, NGOs, and People's Organizations in Kenya, Tanzania, and Uganda*. Portsmouth, NH: Heinemann, 1995.

Sengendo, Hannington, Tonny Oyana, Bob Nakuleza, and Paul Musali. "The Informal Sector in Employment Creation in Kampala." In *Negotiating Social Space: East African Microenterprises*, edited by Patrick Alila and Poul Pedersen, 25–50. Trenton, NJ: Africa World Press, 2001.

Serres, Michel. *Malfeasance: Appropriation through Pollution*. Stanford, CA: Stanford University Press, 2010.

———. *The Parasite*. Minneapolis: University of Minnesota Press, 2007.

Singh, Nikhil. *Black Is a Country: Race and the Unfinished Struggle for Democracy*. Cambridge, MA: Harvard University Press, 2004.

Sjogren, Anders. *Between Militarism and Technocratic Governance: State Formation in Contemporary Uganda*. Kampala: Fountain Publishers, 2013.

Smith, James Howard. *Bewitching Development: Witchcraft and the Reinvention of Development in Neoliberal Kenya*. Chicago: University of Chicago Press, 2008.

Solomon, Marisa. "'The Ghetto Is a Gold Mine': The Racialized Temporality of Betterment." *International Labor and Working-Class History* 95 (2019): 76–94.

Southall, Aidan, and Peter Gutkind. *Townsmen in the Making: Kampala and Suburbs*. Kampala: East Africa Institute, 1957.

Stallybrass, Peter. "Marx and Heterogeneity: Thinking the Lumpenproletariat." *Representations*, 1990, 69–95.

Stallybrass, Peter, and Allison White. *The Politics and Poetics of Transgression*. Ithaca, NY: Cornell University Press, 1986.

Stamatopoulou-Robbins, Sophia. "Failure to Build: Sewage and the Choppy Temporality of Infrastructure in Palestine." *Environment and Planning E*, 2020.

———. *Waste Siege: The Life of Infrastructure in Palestine.* Stanford, CA: Stanford University Press, 2019.

Star, Susan Leigh. "The Ethnography of Infrastructure." *American Behavioral Scientist* 43, no. 3 (1999): 377–91.

Stoler, Ann Laura. *Carnal Knowledge and Imperial Power: Race and the Intimate in Colonial Rule.* Berkeley: University of California Press, 2002.

———. *Race and the Education of Desire: Foucault's History of Sexuality and the Colonial Order of Things.* Durham, NC: Duke University Press, 1995.

Strasser, Susan. *Waste and Want: A Social History of Trash.* New York: Norton, 2000.

Street, Alice. *Biomedicine in an Unstable Place: Infrastructure and Personhood in a Papua New Guinea Hospital.* Durham, NC: Duke University Press, 2014.

Summers, Carol. "Intimate Colonialism: The Imperial Production of Reproduction in Uganda,1907–1925." *Signs* 16, no. 4 (1991): 787–807.

———. "Radical Rudeness: Ugandan Social Critiques during the 1940s." *Journal of Social History* 39, no. 3 (2006): 741–70.

Swanson, Maynard. "The Sanitation Syndrome: Bubonic Plaque and Urban Native Policy in the Cape Colony, 1900–1909." *Journal of African History* 18, no. 3 (1977): 287–410.

Talbott, Taylor. "A Green Army Is Ready to Keep Plastic Waste out of the Ocean." *Scientific American Blog Network*, October 7, 2019. https://blogs .scientificamerican.com/observations/a-green-army-is-ready-to-keep-plastic -waste-out-of-the-ocean.

Taussig, Michael. *Defacement: Public Secrecy and the Labor of the Negative.* Stanford, CA: Stanford University Press, 1999.

Taylor, Dorceta. *Toxic Communities: Environmental Racism, Industrial Pollution, and Residential Mobility.* New York: New York University Press, 2014.

Taylor, Edgar. "Claiming Kabale: Racial Thought and Urban Governance in Uganda." *Journal of East African Studies* 7, no. 1 (2013): 143–63.

Taylor, Keeanga-Yamahtta. "Five Years Later, Do Black Lives Matter?" *Jacobin Magazine*, September 30, 2019. https://jacobinmag.com/2019/09/black-lives -matter-laquan-mcdonald-mike-brown-eric-garner.

———. "Predatory Inclusion." *N+1*, September 24, 2019. https://nplusonemag.com /issue-35/essays/predatory-inclusion.

Thiong'o, Ngũgĩ wa. *Decolonising the Mind: The Politics of Language in African Literature.* London: James Currey, 1986.

Tiberondwa, Ado. *Missionary Teachers as Agents of Colonialism in Uganda.* Kampala: Fountain Publishers, 1977.

Ticktin, Miriam. *Casualties of Care: Immigration and the Politics of Humanitarianism in France.* Berkeley: University of California Press, 2011.

Transparency International Uganda. *Up against Giants: Oil-Influenced Land Injustices in the Albertine Graben in Uganda.* Kampala: Transparency International Uganda, 2015. http://creduganda.org/wp-content/uploads /2019/05/up-against-giants.pdf.

Tripp, Aili Mari. *Museveni's Uganda: Paradoxes of Power in a Hybrid Regime.* Boulder, CO: Lynne Rienner, 2010.

Tsing, Anna Lowenhaupt. *Friction: An Ethnography of Global Connection.* Durham, NC: Duke University Press, 2005.

———. "Inside the Economy of Appearances." *Public Culture* 12, no. 1 (2000): 115–44.

Tveit, Marta. "The Afropolitan Must Go." *Africa Is a Country* (blog), November 28, 2013. https://africasacountry.com/2013/11/the-afropolitan-must-go.

Twaddle, Michael, ed. *Expulsion of a Minority: Essays on Uganda Asians.* London: Athlone Press, 1975.

Uganda Media Center. "President Says KCCA Executive Director Came to Rescue the City from Potholes." Statehouse of Uganda, February 29, 2016. https://www.statehouse.go.ug/media/news/2016/02/29/president-says-kcca -executive-director-came-rescue-city-potholes-flies-and-sha.

Vasudevan, Pavithra. "An Intimate Inventory of Race and Waste." *Antipode,* 2019.

Vaughan, Megan. *Curing Their Ills: Colonial Power and African Illness.* Stanford, CA: Stanford University Press, 1991.

Vermeiren, Karolien, Anton Van Rompaey, Maarten Loopmans, Eria Serwajja, and Paul Mukwaya. "Urban Growth of Kampala, Uganda: Pattern Analysis and Scenario Development." *Landscape and Urban Planning* 106, no. 2 (2012): 199–206.

Vinsel, Lee, and Andrew Russell. "Hail the Maintainers." *Aeon* (blog). https:// aeon.co/essays/innovation-is-overvalued-maintenance-often-matters-more.

Von Schnitzler, Antina. *Democracy's Infrastructure: Techno-Politics and Protest after Apartheid.* Princeton, NJ: Princeton University Press, 2016.

Voyles, Traci Brynne. *Wastelanding: Legacies of Uranium Mining in Navajo Country.* Minneapolis: University of Minnesota Press, 2015.

Wallman, Sandra. *Kampala Women Getting By: Wellbeing in the Time of AIDS.* Athens, OH: James Currey, Fountain Publishers, Ohio University Press, 1996.

Watson, Vanessa. "African Urban Fantasies: Dreams or Nightmares?" *Environment and Urbanization* 26, no. 1 (2014): 215–31.

———. "Seeing from the South: Refocusing Urban Planning on the Globe's Central Urban Issues." *Urban Studies* 46, no. 11 (2009): 2259–75.

Weheliye, Alexander. *Habeas Viscus: Racializing Assemblages, Biopolitics, and Black Feminist Theories of the Human.* Durham, NC: Duke University Press, 2014.

Weiss, Brad. *Street Dreams and Hip Hop Barbershops: Global Fantasy in Urban Tanzania*. Bloomington: Indiana University Press, 2009.

West, Harry, and Todd Sanders, eds. *Transparency and Conspiracy: Ethnographies of Suspicion in the New World Order*. Durham, NC: Duke University Press, 2003.

West, Michael. *The Rise of an African Middle Class: Colonial Zimbabwe, 1898–1965*. Bloomington: Indiana University Press, 2002.

White, E. Frances. *Dark Continent of Our Bodies: Black Feminism and the Politics of Respectability*. Philadelphia: Temple University Press, 2001.

White, Luise. *The Comforts of Home: Prostitution in Colonial Nairobi*. Chicago: University of Chicago Press, 1990.

Whitehouse, G. C. "The Building of the Kenya and Uganda Railway." *Uganda Journal* 12, no. 1 (1948): 1–15.

Whitson, Risa. "Negotiating Place and Value: Geographies of Waste and Scavenging in Buenos Aires." *Antipode* 43, no. 4 (2011): 1404–33.

Whyte, Susan Reynolds, ed. *Second Chances: Surviving AIDS in Uganda*. Durham, NC: Duke University Press, 2014.

Whyte, Susan Reynolds, Michael Whyte, Lotte Meinert, and Jenipher Twebaze. "Therapeutic Citizenship: Belonging in Uganda's Projectified Landscape of AIDS Care." In *When People Come First: Critical Studies in Global Health*, edited by João Biehl and Adriana Petryna, 140–65. Princeton, NJ: Princeton University Press, 2013.

Williams, Raymond. *The Long Revolution*. Orchard Park, NY: Broadview Press, 2001.

———. *Marxism and Literature*. New York: Oxford University Press, 1977.

Winner, Langdon. "Do Artifacts Have Politics?" *Daedalus* 109, no. 1 (1980): 121–36.

Wolcott, Victoria. *Remaking Respectability: African American Women in Interwar Detroit*. Chapel Hill: University of North Carolina Press, 2001.

Woolgar, Steven. "Configuring the User: The Case of Usability Trials." In *A Sociology of Monsters: Essays on Power, Technology, and Domination*, edited by John Law, 58–102. New York: Routledge, 1991.

World Bank. "The Growth Challenge: Can Ugandan Cities Get to Work? Uganda Economic Update, 5th Edition." Washington, DC.: World Bank, 2015. https://documents.worldbank.org/en/publication/documents-reports/documentdetail/145801468306254958/the-growth-challenge-can-ugandan-cities-get-to-work.

———. "World Bank Open Data." https://data.worldbank.org.

Wyrod, Robert. *AIDS and Masculinity in the African City*. Berkeley: University of California Press, 2016.

Young, Graeme. "From Protection to Repression: The Politics of Street Vending in Kampala." *Journal of Eastern African Studies* 11, no. 4 (2017): 714–33.

Zimmerman, Andrew. *Alabama in Africa: Booker T. Washington, the German Empire, and the Globalization of the New South*. Princeton, NJ: Princeton University Press, 2012.

Žižek, Slavoj. "Ecology." In *Examined Life: Excursions with Contemporary Thinkers*, edited by Astra Taylor, 155–84. New York: New Press, 2009.

Zoanni, Tyler. "Appearances of Disability and Christianity in Uganda." *Cultural Anthropology* 34, no. 3 (2019): 444–70.

Index

Founded in 1893,
UNIVERSITY OF CALIFORNIA PRESS
publishes bold, progressive books and journals
on topics in the arts, humanities, social sciences,
and natural sciences—with a focus on social
justice issues—that inspire thought and action
among readers worldwide.

The UC PRESS FOUNDATION
raises funds to uphold the press's vital role
as an independent, nonprofit publisher, and
receives philanthropic support from a wide
range of individuals and institutions—and from
committed readers like you. To learn more, visit
ucpress.edu/supportus.